INTRODUCTION TO DISJUNCTIVE KRIGING AND NON-LINEAR GEOSTATISTICS

INTRODUCTION TO

DISJUNCTIVE KRIGING

AND

NON-LINEAR GEOSTATISTICS

Jacques Rivoirard

CLARENDON PRESS · OXFORD
1994

Oxford University Press, Walton Street, Oxford OX2 6DP
Oxford New York
Athens Auckland Bangkok Bombay
Calcutta Cape Town Dar es Salaam Delhi
Florence Hong Kong Istanbul Karachi
Kuala Lumpur Madras Madrid Melbourne
Mexico City Nairobi Paris Singapore
Taipei Tokyo Toronto
and associated companies in
Berlin Ibadan

Oxford is a trade mark of Oxford University Press

Published in the United States by
Oxford University Press Inc., New York

British Library Cataloguing in Publication Data
Data available

Library of Congress Cataloging in Publication Data
Data available
ISBN 0 19 874180 4

1 3 5 7 9 10 8 6 4 2

Printed in Great Britain
on acid-free paper by
Biddles Ltd
Guildford and King's Lynn

Preface

This book gives a comprehensive view of non-linear geostatistics, from the first work by G. Matheron and A. Maréchal in the 1970s, to the latest developments related to geometric problems.

The course notes, which form the first part of the book, correspond to the course on non-linear geostatistics for the CFSG, a postgraduate program in geostatistics offered by the Centre de Géostatistique de Fontainebleau (Ecole des Mines de Paris).

In the second part of the book, three case-studies illustrate the subject.

I am very grateful to the geologists and engineers of COGEMA who have participated in the developments on geometric problems and have made the first case-study possible.

I would also like to thank my colleagues Margaret Armstrong and Richard Webster for having supported and translated this work, and for the two other case-studies.

Contents

Usual notations

Y, Z	variables
x	a point in space
$Z(x)$	variable (grade for instance) at point x
$Y(x)$	another variable at point x (e.g. gaussian transform of the grade)
x_α, x_β	sample points
Z_α	$Z(x_\alpha)$
v	a block, or its support
V	a larger support
$Z(v)$	grade on support v
Y_v	gaussian transform of $Z(v)$
$E(Z)$	expectation (mean value) of Z
$\text{var}\ Z$	variance of Z
$\text{cov}\ (Y, Z)$	covariance between Y and Z
h	a distance, or a distance vector
$C(h)$	covariance at distance h
$\gamma(h)$	variogram
Sph (20)	spherical variogram with range 20 and sill 1
$F(z)$	cumulative distribution function of $Z(x)$ $= P(Z(x) < z)$ probability of having $Z(x) < z$
$F_v(z)$	cumulative distribution function of $Z(v)$ $= P(Z(v) < z)$ probability of having $Z(v) < z$
$I_{Y(x) \geq j}$	or $I(Y(x) \geq j)$, indicator of $Y(x) \geq j$
T_j	$P(Y(x) \geq j)$
T_{ij}^h	$P(Y(x) \geq i \quad Y(x+h) \geq j)$
$\gamma_j(h)$	variogram of indicator $I(Y(x) \geq j)$
$\gamma_{ij}(h)$	cross-variogram between indicators $I(Y(x) \geq i)$ and $I(Y(x) \geq j)$
z_c	cut-off for $Z(x)$ or $Z(v)$
y_c	cut-off for gaussian variable $Y(x)$ or Y_v
$g(t)$	probability distribution function of a standard gaussian
$G(t)$	cumulative distribution function of a standard gaussian
$H_n(y)$	factor of order n (in gaussian models: Hermite polynomial of degree n)
Y_α	$Y(x_\alpha)$
$H_n(Y_\alpha)$	$H_n[Y(x_\alpha)]$
*	as superscript implies an estimator
K	as superscript implies kriging estimator
SK	simple kriging

OK	ordinary kriging
CK	cokriging
DK	disjunctive kriging
CE	conditional expectation
λ_α	weight of sample point x_α
$\lambda_{n\alpha}$	weight of Hermite polynomial $H_n(Y_\alpha)$ at sample point x_α
ρ	a correlation coefficient
$\rho(h)$	a correlogram (normed covariance) at distance h
$\rho_{\alpha\beta}$	correlation between $Y(x_\alpha)$ and $Y(x_\beta)$
$\Phi(y)$	gaussian anamorphosis function of $Z(x)$
φ_n	coefficient of order n when expanding Φ into Hermite polynomials
σ^2	a variance
σ_K^2	kriging variance
Φ_v	gaussian anamorphosis function of $Z(v)$
r	change of support coefficient
T	proportion of ore
Q	quantity of metal
B	conventional profit

Introduction

Why such a book?

The discipline that we now know as geostatistics began its development more than thirty years ago to tackle problems posed in evaluating bodies of ore in mines, and it has been extended since to other fields of activity. In particular it was around 1960 that G. Matheron propounded linear geostatistics, which embodied certain tools, such as the variogram and kriging, now in everyday use. These tools enable us to evaluate, for example, the quantity of metal in a mining block or panel, by estimating the mean grade of such blocks from values measured on samples. Such mining applications are described in David (1977) and Journel and Huijbregts (1978). Webster and Oliver (1990) give a presentation of these linear tools for soil scientists. We should note in passing the existence of non-stationary geostatistics that takes into account drift, as in the roof of a petroleum reservoir. The program *Bluepack* was written for this.

Although linear geostatistics is not the subject of this book we may note that knowledge of it often stops at the variogram and kriging, and that many practitioners neglect to calculate the precision of global estimates when using kriging for local estimation. In future practitioners should take more interest in linear methods for global estimation and in transitive methods in particular. This should not be difficult because there are standard works by Matheron (1965, 1971) and by Journel and Huijbregts (1978) on the subject.

The situation is different with non-linear geostatistics. It, together with the parameterization of reserves, is one of the least understood branches of geostatistics. There are two reasons for this: the lack of any comprehensive book on the subject, and the mathematical nature of the publications that do exist.

Beginning with the second point, it is true that the mathematics and its arsenal needed for non-linear geostatistics are more complex than those of linear geostatistics, and that the user may be put off by their presentation. Nevertheless, the user must master a minimum of the mathematics if he is to use the methods, and in particular he must understand why and how the mathematics plays its part.

Let us take the example of lognormal kriging. Here one kriges not the measured variable itself but its logarithm, which is assumed to be normally distributed. In order to obtain a final estimate on the original scale it is not sufficient simply to take the exponent of the kriged value. We can imagine the astonishment of some readers and, perhaps, the anguish of others. Exponentiation alone introduces a bias. How can we explain that and obtain the correct form for the estimator if we do not understand enough of the theory?

We mentioned above that there was no comprehensive work on non-linear geostatistics. Not only that, the subject itself is by no means comprehensive: only a few tools and models in this branch of the subject are used at present (for example, models with transformations to gaussian form including log-transformation, and change of support using the discrete gaussian model). For many people DK (disjunctive kriging) is just bigaussian DK. Other models have been used, but only rarely, even though they are better suited in some circumstances than the ones that have been used. Their everyday use is something for the future. This does not mean that theory already provides all the tools needed: in particular, there is no theory for non-linear non-stationary geostatistics yet. The theoretical obstacle resides in modelling simultaneously the drift and the residuals from the drift when the differences between them are not known from measurements.

This work is intended principally for the student or practitioner who is already familiar with linear geostatistics and who wishes to extend his knowledge and expertise to tackle non-linear problems. We have tried, as far as possible, to build on the reader's current knowledge and experience of linear geostatistics. The fundamental development is described in detail, but so as not to make the book too heavy going we treat briefly some matters of only secondary importance, such as estimation variances and problems of convergence. Other topics such as DK density or the variogram of order 1 we have omitted altogether. The text is completed with three case-studies that illustrate the ideas. Two are simple applications to mining; the third comes from agriculture. The aim has been not to cover the subject comprehensively, but to provide the reader with a sufficient basic understanding of the theory so that he can pursue the subject further in the more specialized literature.

Why non-linear geostatistics?

By the 1960s non-linear methods based on the lognormal distribution had already been developed to serve the needs of miners in the South African goldfields. To estimate the value in a panel from sample data D. G. Krige had proposed lognormal regression, which later led him to lognormal kriging and which is still much used today. Before that H. S. Sichel had been concerned to evaluate deep ore bodies from few sample data. He had proposed an estimator and provided confidence intervals for the mean, assuming that the sample values were independent of one another and distributed lognormally.

It was only in the 1970s, however, that non-linear technique proper came into being. Among the increasingly complex problems faced by miners was that of forecasting the results of selective exploitation. To simplify, let us imagine that before exploitation a deposit consists of numerous small selection blocks, all identical and of the same weight. Fig. 0.1 represents schematically a horizontal section through such a deposit.

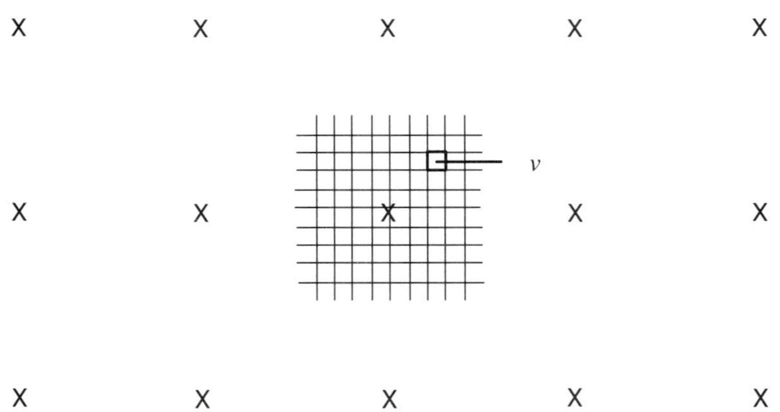

Fig. 0.1

Before beginning to extract the ore one knows only the grades of the survey samples, whose positions are marked by crosses in Fig. 0.1. It is then quite possible to estimate, either by kriging or some other method, the grade, $Z(v)$, of each small block v. This estimate will rarely be precise, however. The fact that the grade of a block is estimated at 1%, for example, does not mean that this grade, as it will be known at the time of extraction, cannot exceed a threshold of 2%. When one seeks to forecast the recoverable reserves of small blocks it is therefore not the grade of the block, $Z(v)$, that one must estimate, but two other quantities.

One is the indicator

$$I_{Z(v)>2}$$

(i.e. the variable that equals 1 if $Z(v)>2$ and 0 otherwise, also denoted $I(Z(v) > 2)$), which when multiplied by the weight of the block gives the quantity of the ore recovered at the threshold of 2%.

The other is the variable $Z(v)\ I(Z(v) > 2)$, the product of the grade and the indicator which gives the quantity of metal that will be recovered at 2%. Thus, we have a complex problem that cannot be solved using linear combinations of the grades alone, and it is for this reason that we need non-linear geostatistical techniques.

Let us continue with our example. At the time of prediction one is not in general concerned with estimating the recoverable reserve of either ore or metal from each small block individually. Rather one wants an estimate of the total reserve of all the blocks within a mining panel or even all the blocks in the deposit. It will no longer be the indicator of the block that we estimate, but the mean of the indicators in all N blocks:

$$\frac{1}{N} \sum_i I_{Z(v_i)>2}$$

that is the proportion of blocks the grade of which exceeds the threshold.

When one wants to estimate global reserves (and is therefore concerned with the whole deposit) the problem becomes again one of knowing the statistical distribution of the block grades. Of this distribution we know only the mean, which is the same as that of the sample, and the variance, which we can calculate using linear geostatistics. To go from the distribution of the samples on quasi-punctual supports to that of the blocks necessitates a model for the change of support, which is described in Chapter 10. Naturally, such a change of support will be equally necessary for local estimation of one or several blocks, and this is dealt with in Chapters 11 and 12.

In cases where there is no change of support to take into account we may wish to estimate, for example, the indicator:

$$I_{Z(x)>2}$$

at a point x or the mean (regularized value) of the indicator over a panel:

$$\frac{1}{V} \int_V I_{Z(x)>2} \, dx$$

that is the proportion of the panel occupied by points whose values exceed 2. This is a matter of punctual estimation, which is dealt with in Chapters 1 to 9. In the simplest cases (Chapter 3) $I(Z(x) > 2)$ will be well estimated by kriging from the same indicator defined at the sampling points: $I(Z(x_\alpha) > 2)$. In general, however, more elaborate algorithms are needed (Chapters 4, 5, 7, and 9).

The result of estimating an indicator, a variable that takes only the values 0 and 1, at a point x is in a sense a probability: here the probability that the true value at that point exceeds the threshold 2. This is best regarded as only a pseudo-probability, however, because the result can lie outside the interval [0,1].

As above, the methods outlined have been mainly used in mining. Nevertheless, they have application in other domains, and there the interest is not so much the estimation of the measured variable, which is often imprecise, but the probability with which unknown values exceed or fall short of some critical threshold. In environmental monitoring, for example, one may be concerned that a pollutant does not exceed some defined concentration, and in agriculture a farmer or grazier may wish to ensure that some minimum concentration of nutrient in the soil (a deficiency threshold) is exceeded (see the second case-study). So although the book emphasizes mining, it is addressed to a much wider readership. For this reason some points that are specific to mining, for example the information effect, are played down, and the examples schematized, so that other readers will draw profit from the book. In other respects the variables that a miner has to deal with such as the grade of an ore are readily matched with those of concern to an environmental or agricultural scientist, such as the concentration of a pollutant or nutrient.

Finally let us note that non-linear techniques can be used instead of linear ones for straightforward estimation of the measured variable, as in lognormal kriging.

PART ONE: COURSE NOTES

1

Reminder on cokriging

Cokriging will be used in the following chapters to define the disjunctive kriging estimator. In the present chapter we give some reminders, in particular of the cases where cokriging simplifies.

Why use cokriging?

Should we use cokriging to estimate several variables or would it suffice to krige them separately? There are two main reasons for using cokriging.

(1) When kriging an undersampled variable, cokriging allows us to take advantage of the information available on another variable which is related to the first and for which there are more data.

(2) When the values of all variables are available at all sample points, it improves the coherence between the estimated values by taking account of the relationship between the variables.

We now consider the second case. As a simple example we shall estimate the bottom $Z_1(x)$ and the top $Z_2(x)$ of a seam of rock. Kriging gives the best linear estimate of each of the variables separately. But most of the time we are also interested in the thickness of the seam $T(x)$:

$$T(x) = Z_2(x) - Z_1(x).$$

Knowing any two of the three variables $Z_1(x)$, $Z_2(x)$, and $T(x)$ we can obtain the third one by simple addition or subtraction. Unfortunately the same simple relations do not apply to the kriged estimates. In general the difference:

$$Z_2(x)^K - Z_1(x)^K$$

does not equal $T(x)^K$.

In contrast, cokriging gives the best linear estimate of the variables and of linear combinations of them. So the cokriged estimates satisfy:

$$Z_2(x)^{CK} - Z_1(x)^{CK} = T(x)^{CK}.$$

Comments

Kriging a positive variable does not guarantee that the resulting estimate will be positive. Similarly cokriging the above variables does not guarantee that the cokriged estimate of thickness $T(x)^{CK}$ will be positive, either. To ensure this, more elaborate techniques are required.

Simplifying cokriging

In order to use cokriging, we need to know the cross-variograms (or cross-covariances) of the variables as well as the auto-variograms. A mathematically consistent coregionalization model must be fitted to all these variograms, whose number increases with the square of the number of variables. The size of the cokriging system also increases with the number of variables. So it will be profitable if cokriging can be simplified.

In fact, two types of model allow us to do this. So it is useful in practice to try to get back to these. In both cases the cokriging estimates are simply obtained by solving some kriging systems.

In what follows we shall assume stationarity unless explicitly stated otherwise.

Intrinsically correlated variables

Let $Z_1(x)$, $Z_2(x)$, . . . $Z_n(x)$ be variables whose auto- and cross-covariances (or only variograms) are all the same apart from a multiplicative factor. That is, if the cross-covariance between $Z_i(x)$ and $Z_j(x)$ is $C_{ij}(h)$, then:

$$C_{ij}(h) = \text{cov}\,[Z_i(x),\ Z_j(x+h)] = C_{ij}(0)\,\rho(h)$$

where $\rho(h)$ is a spatial correlation function and $C_{ij}(0)$ is a constant.

In general the correlation between two variables $Z_i(v)$ and $Z_j(v)$ depends on their support v (which may be just a point x) and on the field V described by v. It is not an inherent characteristic. On the contrary, when all the variograms are multiples of each other the correlation is said to be intrinsic: it does not depend on v and V (Matheron 1965: 149).

Using this model (and assuming that all the variables have been measured at all sample points), cokriging any variable (or a linear combination of the variables) is just the same as kriging that variable. Only one kriging system has to be solved, the weights for any sample are the same for all the variables.

Factorizing the Z_1, Z_2, . . . Z_n

Suppose there is no spatial correlation between $Z_1(x)$, $Z_2(x)$, . . . $Z_n(x)$, i.e. their correlation coefficients are zero and their cross-covariances are identically null. Then kriging each of these variables is equivalent to cokriging it.

Suppose now a correlation exists between these variables. It is sometimes possible to go back to other variables which have no spatial correlation and whose kriging will give the cokriging of all variables. This is the case when we can combine linearly the $Z_1(x)$, $Z_2(x)$, . . . $Z_n(x)$ to build a set of spatially uncorrelated variables $Y_i(x)$, which can in their turn be combined linearly to give the $Z_j(x)$:

$$Z_j(x) = \sum_i a_{ji}\ Y_i(x).$$

These variables Y_i, which can be obtained at each sampling point from the Z_j at that point, are spatially uncorrelated; and so kriging them is equivalent to cokriging them:

$$Y_i^{CK} = Y_i^K.$$

Then we obtain the cokriged estimates of the Z_j:

$$Z_j^{CK} = \sum_i a_{ji}\ Y_i^K.$$

We call these variables Y_i factors, and so factorizing the Z_j is simply finding their decomposition into factors.

As there is no spatial correlation between the factors, their estimation errors $Y_i - Y_i^K$ are uncorrelated, and so their estimation variances give that of Z_j:

$$Z_j - Z_j^{CK} = \sum_i a_{ji}\ (Y_i - Y_i^K)$$

$$\text{var}\ [Z_j - Z_j^{CK}] = \sum_i (a_{ji})^2\ \text{var}\ (Y_i - Y_i^K).$$

Examples

Let us now return to our initial example to illustrate these models.

(a) Assume that the top $Z_2(x)$ and the bottom $Z_1(x)$ of a seam have the following coregionalization model:

$$C_1(h) = 16 \text{ Sph } (100) + 16 \text{ Sph } (20)$$
$$C_2(h) = 25 \text{ Sph } (100) + 25 \text{ Sph } (20)$$
$$C_{12}(h) = 16 \text{ Sph } (100) + 16 \text{ Sph } (20)$$

where $C_1(h)$, $C_2(h)$, and $C_{12}(h)$ denote the auto- and cross-covariances for the variables, and where 16 Sph (100) denotes a spherical model with a range of 100 and a sill of 16. These three covariance functions are all proportional to Sph (100) + Sph (20). And so Z_1 and Z_2 are intrinsically correlated.

(b) Now consider this coregionalization model:

$$C_1(h) = 16 \text{ Sph } (100)$$
$$C_2(h) = 25 \text{ Sph } (100) + 25 \text{ Sph } (20)$$
$$C_{12}(h) = 16 \text{ Sph } (100).$$

Here the structure of Z_1, $C_1(h)$, is the same as the cross-structure, $C_{12}(h)$. This enables us to rewrite the structure of Z_2 as:

$$C_2(h) = 16 \text{ Sph } (100) + 9 \text{ Sph } (100) + 25 \text{ Sph } (20).$$

This means that we can represent $Z_2(x)$ as the sum of two components, $Z_1(x)$ and $T(x)$:

$$Z_2(x) = Z_1(x) + T(x)$$

where $T(x) = Z_2(x) - Z_1(x)$ has the covariance:

$$9 \text{ Sph } (100) + 25 \text{ Sph } (20)$$

and is spatially uncorrelated with $Z_1(x)$. The model has effectively been factorized into $Z_1(x)$ and $T(x)$, and so we can krige these components separately to obtain the cokriged estimates:

$$Z_1(x)^{CK} = Z_1(x)^K$$
$$T(x)^{CK} = T(x)^K$$
$$Z_2(x)^{CK} = Z_1(x)^{CK} + T(x)^{CK}$$
$$= Z_1(x)^K + T(x)^K.$$

Comments on this last example

1. Orthogonality

As we are dealing with stationary variables we can calculate their means and then write them as:

$$Z_1(x) = m_1 + (Z_1(x) - m_1)$$

$$Z_2(x) = m_2 + (Z_1(x) - m_1) + (T(x) - m_T)$$

with $m_2 = m_1 + m_T$. We have:

$$\text{cov}\,[Z_1(x),\ T(x+h)] = E[(Z_1(x) - m_1)(T(x+h) - m_T)] = 0.$$

Since the expected value of the product of $Z_1(x) - m_1$ and $T(x+h) - m_T$ is zero for any x and h, the two random functions $Z_1(x) - m_1$ and $T(x) - m_T$ are said to be orthogonal.

2. Nested structures

The variable $T(x)$, which has the covariance:

$$9 \text{ Sph } (100) + 9 \text{ Sph } (20)$$

can be split into two spatially uncorrelated components with ranges of 100 and 20 respectively. But, in contrast to $Z_1(x)$ and $T(x)$, these components cannot be obtained experimentally at a given point x from $Z_1(x)$ and $Z_2(x)$.

Exercise: factorization of a product

In some cases we have to consider the joint estimation of two variables $Z(x)$ and $T(x)$ and of their product $Q(x) = Z(x)T(x)$.

Suppose for instance that the variable $Z(x)$ to be estimated is defined on a domain the geometry of which is not precisely known. Then it is worth considering also the estimation of the indicator (0 or 1 variable) $T(x)$ of this domain. Moreover it is convenient to introduce the variable $Q(x) = Z(x)T(x)$ which is extended to 0 outside the field of $Z(x)$.

Another instance: in 2D $T(x)$ may represent the thickness of a mineralized vein, that is, the length of its crossing at point x, and $Z(x)$ the grade in metal of this crossing. Then $Q(x)$ is the metal accumulation of this crossing. For a block, the average thickness is equal to the arithmetic average of the thicknesses of its different points, and the same for the average accumulation. The average grade of the block (metal/ore):

$$\frac{\int_V Q(x)\,dx}{\int_V T(x)\,dx} = \frac{\frac{1}{V}\int_V Q(x)\,dx}{\frac{1}{V}\int_V T(x)\,dx} = \frac{\frac{1}{V}\int_V Z(x)\,T(x)\,dx}{\frac{1}{V}\int_V T(x)\,dx}$$

is the average of the grades, weighted by their thickness (a large grade gives the block more metal if it corresponds to a large thickness). Unlike the grades, which have to be weighted, the variables thickness and accumulation are additive and suited to linear estimation.

We suppose first that $Z(x)$ and $T(x)$ are two stationary and independent random functions (then they have no spatial correlation).

(1) We know $T(x)$ at a point x. Show that the regression (conditional expectation) of $Q(x)$ at this point can be written:

$$E[Q(x)\,|\,T(x)] = m\,T(x)$$

with $m = E[Z(x)]$.

(2) Verify that the residual of this regression:

$$R(x) = Q(x) - E[Q(x)\,|\,T(x)]$$
$$= [Z(x) - m]\,T(x)$$

has no correlation with $T(x)$ at the same point x. Show that $R(x)$ has no correlation either with $T(x+h)$ at any other point $x+h$.

(3) We can split $Q(x)$

$$Q(x) = m\,T(x) + R(x)$$

where $T(x)$ and $R(x)$ are spatially uncorrelated. Deduce that the cokriging of $T(x)$ and $Q(x)$ can be obtained by kriging separately the factors $T(x)$ and $R(x)$:

$$T(x)^{CK} = T(x)^{K}$$
$$Q(x)^{CK} = m\,T(x)^{CK} + R(x)^{CK}$$
$$= m\,T(x)^{K} + R(x)^{K}.$$

(4) Suppose that $Z(x)$ has a pure nugget structure. Show that $R(x)$ is also pure nugget. Its kriging at an unknown point x equals its mean,

i.e. 0. Deduce the cokriging of $Q(x)$. Show that the estimation of $Z(x)$ which is obtained as the ratio between the cokrigings of $Q(x)$ and $T(x)$ is constant. Is it consistent with the nugget structure of the grades?

(5) Show that the cross-covariance (or cross-variogram) between $T(x)$ and $Q(x)$ is proportional to the simple structure of $T(x)$.

(6) Independence between $T(x)$ and $Z(x)$ is a strong hypothesis. Show that the factorization through $T(x)$ and $R(x)$ holds when the variables $T(x)$ and $Q(x)$ are only independent 'on average':

$$E[Z(x) \mid T(x),\ T(x+h)] = m \qquad \forall h.$$

In practice this may really be a more realistic hypothesis. In the case of a vein for instance, grades and thicknesses are rarely independent, because we expect the grades of small crossings to take much smaller as well as much larger values than for large crossings, even if on average, the grades are the same, for small or large crossings.

2

Indicators and disjunctive kriging

Linear geostatistics deals only with linear combinations of the variable under study $Y(x)$. This makes it possible to estimate linearly the value of the variable at a given point or over a given domain.

Sometimes we need to estimate, not $Y(x)$ itself, but one or more functions $f[Y(x)]$ of $Y(x)$. Considering linear combinations of the variable $Y(x)$ is no longer sufficient and we call on non-linear methods: in particular the conditional expectation that will be given in Chapter 7, and disjunctive kriging introduced in the present chapter.

What makes the indicators interesting is the fact that any function $f[Y(x)]$ can be expressed in terms of the indicators of $Y(x)$. By cokriging these indicators we obtain the DK estimate.

Coding through indicators

Suppose that a random function $Y(x)$ can take a finite number of values. If, for example, these values are 0, 1, 2, and 3, then at each point x, $Y(x)$ must take just one of them. That is,

$$Y(x) = 0, \text{ or } 1, \text{ or } 2, \text{ or } 3.$$

We may code these different possibilities by using indicator functions (called indicators, for short). If $Y(x)=i$ then the indicator $I(Y(x) = i)$ is equal to 1, and to 0 otherwise. Since $Y(x)$ can take only one value at any one place, this coding is said to be disjunctive. At any point x, one and only one of the indicators is equal to 1, and the others are all zero. So:

$$\sum_i I_{Y(x)=i} = I_{Y(x)=0} + I_{Y(x)=1} + I_{Y(x)=2} + I_{Y(x)=3} = 1.$$

Thus we need to know only three of the four indicators in order to deduce the other one.

Developing functions

This disjunctive coding and these indicators are valuable because they allow us to express any function of $Y(x)$ conveniently. Let us consider, for example, a function $f[Y(x)]$ that takes the values f_0, f_1, f_2, and f_3 when $Y(x)$ takes the values 0, 1, 2, and 3. This function can be written as:

$$\sum_i f_i \, I_{Y(x)=i} = f_0 \, I_{Y(x)=0} + f_1 \, I_{Y(x)=1} + f_2 \, I_{Y(x)=2} + f_3 \, I_{Y(x)=3}.$$

We can see this by examining what happens when $Y(x)$ takes each value. If $Y(x)=0$ then $I(Y(x) = 0) = 1$ and the other three indicators are zero. So:

$$\sum f_i \, I_{Y(x)=i} = f_0 \times 1 + f_1 \times 0 + f_2 \times 0 + f_3 \times 0 = f_0.$$

Similarly for the other three possible values.

Examples

1. The variable $Y(x)$ itself can be written using indicators:

$$Y(x) = 0 \, I_{Y(x)=0} + 1 \, I_{Y(x)=1} + 2 \, I_{Y(x)=2} + 3 \, I_{Y(x)=3}.$$

2. Of course we have:

$$I_{Y(x)=2} = 0 \, I_{Y(x)=0} + 0 \, I_{Y(x)=1} + 1 \, I_{Y(x)=2} + 0 \, I_{Y(x)=3}.$$

3. In geostatistics we often use indicators above a cut-off value. For example, the indicator $I(Y(x) \geq 2)$ takes the value 1 when $Y(x) \geq 2$, and 0 otherwise. So we have:

$$\begin{aligned} I_{Y(x)\geq 2} &= I_{Y(x)=2} + I_{Y(x)=3} \\ &= 0 \, I_{Y(x)=0} + 0 \, I_{Y(x)=1} + 1 \, I_{Y(x)=2} + 1 \, I_{Y(x)=3}. \end{aligned}$$

   If $Y(x)$ represents the grade in a particular substance (e.g. metal concentration) of a sample taken at point x, the variable $I(Y(x) \geq 2)$ indicates whether the grade equals or exceeds the cut-off at 2. When multiplied by the weight of the sample this indicator represents the weight of the ore above the cut-off.

4. Similarly the quantity of metal at the 2 cut-off contained in the sample is taken as 0 if the grade $Y(x)$ is less than 2, and is equal to

the product of the grade by the weight of the sample if $Y(x) \geq 2$. So it is equal to:

$$Y(x)\ I_{Y(x) \geq 2}$$

i.e. $$0\ I_{Y(x)=0} + 0\ I_{Y(x)=1} + 2\ I_{Y(x)=2} + 3\ I_{Y(x)=3}$$

multiplied by the weight of the sample.

Limits of linear geostatistics

As the reader knows, linear geostatistics deals only with linear combinations of the variable under study $Y(x)$. This makes it possible to estimate the value of the variable $Y(x)$ at a given point x, or its regularized $(1/V)\int_V Y(x)\,dx$ over a given domain V, using a linear combination of the values $Y(x_\alpha)$ known at sample points x_α (for instance kriging).

Linear geostatistics also provides the variance of estimation (i.e. the variance of the difference between true and estimated values). A small estimation variance means a good precision.

Now let us return to our variable $Y(x)$ which can take the values 0, 1, 2, or 3. Imagine that its kriged estimate $Y(x)^K$ at a given point x equals 1. If the estimation variance is small, there is a high chance for the true value $Y(x)$ not to exceed 2. But if the estimation variance is not small, as frequently happens, the estimated value $Y(x)^K = 1$ only represents an average of tendencies. The true value may be equal to 1, but it may also equal 0, 2, or 3. The fact that the estimate $Y(x)^K$ is smaller than 2 does not mean that the true value $Y(x)$ cannot be equal to or larger than 2, which is represented by the indicator $I(Y(x) \geq 2)$.

This distinction is essential when we want to make predictions relative to cut-off values. In the environmental sciences for instance, we may be interested in estimating in the form of a probability whether a variable $Y(x)$ passes a threshold, e.g. $I(Y(x) \geq 2)$. To predict the results of a selective mining process, we want to estimate indicators such as $I(Y(x) \geq 2)$ (which gives the ore above the cut-off 2), and also the corresponding quantities of metal.

As we can see, computing an estimate for the variable $Y(x)$ is not the solution when we need to estimate, not $Y(x)$ itself, but one or more functions $f[Y(x)]$ of $Y(x)$: indicators, metal quantities, etc. We then call on non-linear methods like the conditional expectation (Chapter 7) or disjunctive kriging.

Disjunctive kriging (DK)

As we have seen, the family of indicators $(I(Y(x) = 0), \ldots, I(Y(x) = 3))$ allows us to express any function of $Y(x)$ as:

$$f(Y(x)) = f_0 \, I_{Y(x)=0} + f_1 \, I_{Y(x)=1} + f_2 \, I_{Y(x)=2} + f_3 \, I_{Y(x)=3}.$$

If we can estimate all of the indicators, we can obtain an estimate of $f[Y(x)]$ for any function f. The disjunctive kriging estimator is defined as that given by cokriging all the indicators:

$$[f(Y(x))]^{DK} = f_0 \, [I_{Y(x)=0}]^{CK} + f_1 \, [I_{Y(x)=1}]^{CK} + \ldots$$

In particular the DK of an indicator is equal to its cokriging with all the indicators:

$$[I_{Y(x)=0}]^{DK} = [I_{Y(x)=0}]^{CK}.$$

Note that disjunctively kriging the sum of two functions is equivalent to summing the two DK estimates (similarly for the difference). That is:

$$[f(Y(x)) + g(Y(x))]^{DK} = [f(Y(x))]^{DK} + [g(Y(x))]^{DK}.$$

This ensures consistency between the estimates of different functions. In particular:

$$[I_{Y(x)\geq i}]^{DK} - [I_{Y(x)\geq i+1}]^{DK} = [I_{Y(x)=i}]^{DK}.$$

Kriging separately the indicators would generally not assure such an equality.

Other expressions for DK

1. Let the x_α denote the data points where the $Y_\alpha = Y(x_\alpha)$ are known. If we develop the cokriging of indicators, we can express $[f(Y(x))]^{DK}$ as a linear combination of the $I(Y(x_\alpha) = i)$, say

$$\sum_\alpha \sum_i f_i^\alpha I_{Y(x_\alpha)=i}$$

where the f_i^α denote the coefficients (the development is tedious although not difficult). But $\Sigma_i \, f_i^\alpha I(Y(x_\alpha) = i)$ defines a function of Y_α (the function f^α which is equal to f_i^α when $Y(x_\alpha) = i$). We then find that the DK of $f[Y(x)]$ is the best estimator of the form $\Sigma_\alpha \, f^\alpha(Y_\alpha)$. This equivalent definition is often used in the specialized literature.

2. In linear kriging of $Y(x)$, we build linear combinations of $Y(x_\alpha)$,

$$\sum_\alpha \lambda_\alpha Y(x_\alpha)$$

with weights λ_α that depend on the variogram, not on the values Y_α themselves. If we want to give to the value Y_α a weight which also depends on Y_α, say $\lambda_\alpha(Y_\alpha)$, we find

$$\sum_\alpha \lambda_\alpha(Y_\alpha) Y_\alpha.$$

As $\lambda_\alpha(Y_\alpha)Y_\alpha$ is a function of Y_α, the best estimator of this form is nothing but DK.

Other generating families

Since in our opening example

$$I_{Y(x)=0} = 1 - \sum_1^3 I_{Y(x)=i}$$

we can write:

$$f(Y(x)) = f_0\ (1 - \sum_1^3 I_{Y(x)=i}) + \sum_1^3 f_i\ I_{Y(x)=i}$$

$$= f_0\ 1 + \sum_1^3 (f_i - f_0)\ I_{Y(x)=i}.$$

This shows that $(1,\ I(Y(x) = 1),\ I(Y(x) = 2),\ I(Y(x) = 3))$ is also a generating family.

Sometimes, it is more convenient to consider the cumulative indicators:

$$\begin{aligned}
I_{Y(x)\geq 0} &= 1 \\
I_{Y(x)\geq 1} &= I_{Y(x)=1} + I_{Y(x)=2} + I_{Y(x)=3} \\
I_{Y(x)\geq 2} &= I_{Y(x)=2} + I_{Y(x)=3} \\
I_{Y(x)\geq 3} &= I_{Y(x)=3}.
\end{aligned}$$

Clearly we have:

$$
\begin{aligned}
I_{Y(x)=3} &= I_{Y(x)\geq 3}\\
I_{Y(x)=2} &= I_{Y(x)\geq 2} - I_{Y(x)\geq 3}\\
I_{Y(x)=1} &= I_{Y(x)\geq 1} - I_{Y(x)\geq 2}\\
I_{Y(x)=0} &= 1 - I_{Y(x)\geq 1}.
\end{aligned}
$$

Since any function of $Y(x)$ can be expressed in terms of the indicators $(I(Y(x) = 0),\ I(Y(x) = 1), I(Y(x) = 2),\ I(Y(x) = 3))$, it can also be written in terms of $(1,\ I(Y(x) \geq 1), \ldots,\ I(Y(x) \geq 3))$, which also forms a generating family.

What do we need for disjunctive kriging?

All the generating families are equivalent. So we may cokrige using $(I(Y(x) = 0), \ldots, I(Y(x) = 3))$ or using $(1,\ I(Y(x) \geq 1), \ldots, I(Y(x) \geq 3))$: let us consider the second case. We clearly must know the auto- and cross-covariance functions (or variograms) of the cumulative indicators.

Let T_i denote the probability that $Y(x)$ exceeds or equals i, which is the expectation of the indicator $I(Y(x) \geq i)$:

$$T_i = P(Y(x) \geq i) = E[I_{Y(x)\geq i}].$$

T_i is just the inverse cumulative distribution function of $Y(x)$. We can write the covariance functions as:

$$
\begin{aligned}
C_1(h) &= \text{cov}\ [I_{Y(x)\geq 1},\ I_{Y(x+h)\geq 1}]\\
&= E[I_{Y(x)\geq 1}\ I_{Y(x+h)\geq 1}] - (E[I_{Y(x)\geq 1}])^2
\end{aligned}
$$

and similarly the cross-covariance functions as:

$$
\begin{aligned}
C_{12}(h) &= \text{cov}\ [I_{Y(x)\geq 1},\ I_{Y(x+h)\geq 2}]\\
&= E[I_{Y(x)\geq 1}\ I_{Y(x+h)\geq 2}] - E[I_{Y(x)\geq 1}]\ E[I_{Y(x+h)\geq 2}].
\end{aligned}
$$

Let T^h_{ij}, where the superscript means at lag h, denote the bivariate distribution of $Y(x)$ and $Y(x + h)$:

$$T^h_{ij} = P(Y(x) \geq i,\ Y(x + h) \geq j) = E[I_{Y(x)\geq i}\ I_{Y(x+h)\geq j}].$$

We have:

$$C_1(h) = T^h_{11} - (T_1)^2$$

$$C_{12}(h) = T^h_{12} - T_1 T_2.$$

This gives:

$$T^h_{11} = C_1(h) + (T_1)^2$$

$$T^h_{12} = C_{12}(h) + T_1 T_2.$$

So knowing the auto- and cross-covariance functions is equivalent to knowing the bivariate distributions of $Y(x)$ and $Y(x + h)$ for any lag h. In the same way that kriging is based on the variogram, so disjunctive kriging is based on these bivariate distributions.

In practice we have to fit a model to the bivariate distributions, or what amounts to the same thing, to fit a model to the coregionalization of the indicators. Using this model makes it possible to perform DK (i.e. cokriging of indicators).

However the size of a cokriging system increases with the number of indicators, here the number of indicator variables. So, as before, we look for models that simplify the cokriging. These are the mosaic model presented in the next chapter and isofactorial models like those we shall see afterwards.

3

The mosaic model: indicators that are intrinsically correlated

The mosaic model is a very simple one in which the auto- and cross-structures of all functions of $Y(x)$ are the same. In consequence the indicator cokriging (i.e. the disjunctive kriging) is reduced to a kriging.

Structure of the model

Suppose we have a region that consists of a mosaic of random compartments in any one of which the value of $Y(x)$ is constant. Moreover, the value in each compartment is independent of the values in the other compartments, but the values all come from a common distribution. An example is given in Fig. 3.1.

Let $\rho(h)$ be the probability that two points, x and $x + h$, a distance h apart, belong to the same compartment. If they do, then $Y(x) = Y(x + h)$. If x and $x + h$ belong to different compartments, which happens with a probability $1 - \rho(h)$, then $Y(x)$ and $Y(x + h)$ are independent. So we have:

$$E[Y(x)\ Y(x+h)] = E[Y(x)^2]\ \rho(h) + [E(Y(x))]^2\ (1 - \rho(h))$$

$$= \rho(h) \operatorname{var} Y(x) + [E(Y(x))]^2.$$

This gives us the covariance:

$$\operatorname{cov}\ [Y(x),\ Y(x+h)] = \rho(h) \operatorname{var} Y(x).$$

Similarly for any function $f(Y(x))$:

$$\operatorname{cov}\ [f(Y(x)),\ f(Y(x+h))] = \rho(h) \operatorname{var} f(Y(x)).$$

If we now consider any two functions f and g of our variable, then it is easy to show that:

$$\operatorname{cov}\ [f(Y(x)),\ f(Y(x+h))] = \rho(h) \operatorname{cov}\ [f(Y(x)),\ g(Y(x))].$$

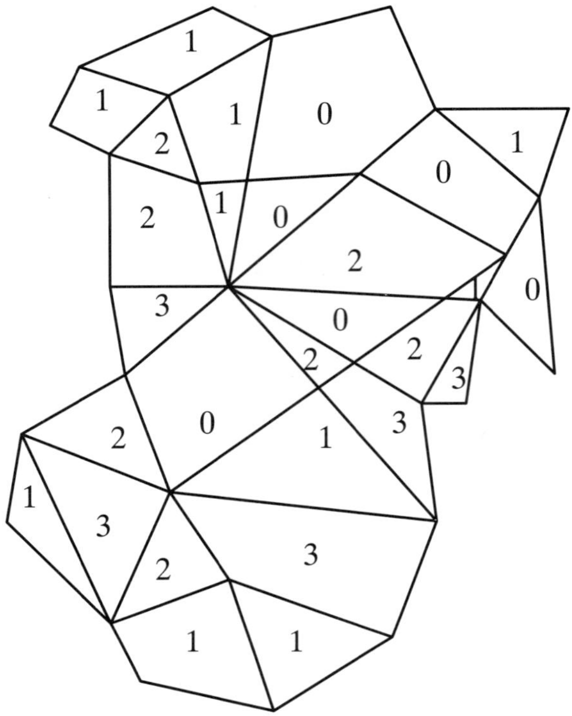

Fig. 3.1 Example of mosaic, with geometrically simple tiles

So the auto- and cross-covariances of all functions of $Y(x)$ are proportional to $\rho(h)$. In particular this is true of the indicator functions, which are therefore intrinsically correlated (see Chapter 1).

DK and indicator kriging

As they are intrinsically correlated cokriging the indicator functions is equivalent to kriging them. We need to solve only a single kriging system, the one based on $\rho(h)$, to get the kriging weights for all the indicator functions, and for all possible functions of $Y(x)$:

$$[f(Y(x))]^{DK} = \sum_i f_i \, [I_{Y(x)=i}]^{CK}$$

$$= \sum_i f_i \, [I_{Y(x)=i}]^{K}$$

$$= [\sum_i f_i \, I_{Y(x)=i}]^K \;=\; [f(Y(x))]^K.$$

As an exercise readers should write out the above krigings explicitly in terms of the kriging weights λ_a, the data $I(Y(x_a)) = i$ and $f(Y(x_a))$, and the mean values $p_i = P[Y(x) = i] = E[I(Y(x)) = i]$ and $E[f(Y(x))]$ which have been assumed to be known. As in current use in geostatistics the x_α denote the sampling points in the neighbourhood of the point x .

Thus the DK of any function is very simple. It is reduced to a kriging, with weights that are the same for all functions. This makes the mosaic model very attractive. This model can be tested experimentally by comparing the auto- and cross-structures of indicators, that must then be identical.

In the models that we are going to consider in the next chapters, the indicators do not all have the same structure. Moreover there always exists some correlation between the indicators. This means that if we just krige the indicators separately the resulting estimate may be far from the optimal one (see Exercise in the next chapter).

Numerical exercise

Imagine that a variable $Y(x)$ takes the values 0, 1, 2, 3, with the probabilities $p_0 = 0.4$, $p_1 = 0.3$, $p_2 = 0.2$, $p_3 = 0.1$. Its mean is

$$E[Y(x)] = 0 \times 0.4 + 1 \times 0.3 + 2 \times 0.2 + 3 \times 0.1 = 1.0.$$

The auto- and cross-covariances of indicators are all proportional to the same $\rho(h) = \exp -|h|/12$: we are in a mosaic model with independent compartment values.

(a) We know the values of $Y(x)$ at the points x_A and x_B of Fig. 3.2 and we want to make estimations at point x_0 using DK. Show that the kriging system with the covariance $\rho(h)$ can be written

$$\lambda_A + \lambda_B \, 0.435 = 0.513$$

$$\lambda_A \, 0.435 + \lambda_B = 0.607$$

and that the solution is $\lambda_A = 0.31$ and $\lambda_B = 0.47$.

(b) The DK of any function $f(Y(x_0))$ (its kriging as we are in the mosaic model) can be written

$$[f(Y(x_0))]^{DK} = \lambda_A f(Y(x_A)) + \lambda_B f(Y(x_B)) + (1 - \lambda_A - \lambda_B) E[f(Y(x))]$$

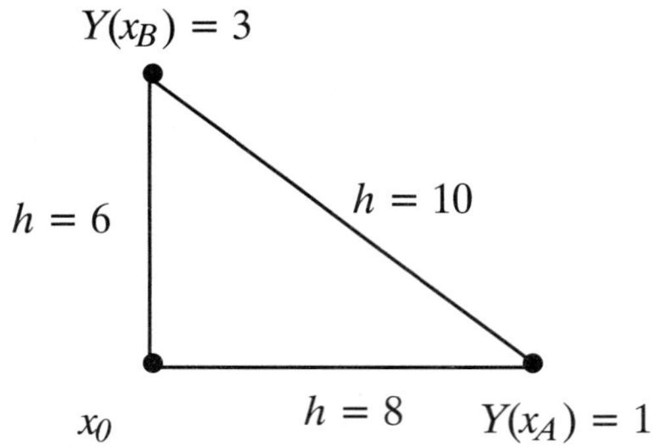

Fig. 3.2

i.e. $$0.31\,f(Y(x_A)) + 0.47\,f(Y(x_B)) + 0.22\ E[f(Y(x))].$$

Deduce the DK estimates:

$$[I_{Y(x_0)=0}]^{DK} = 0.31 \times 0 + 0.47 \times 0 + 0.22 \times 0.4 = 0.09$$

$$[I_{Y(x_0)=1}]^{DK} = 0.31 \times 1 + 0.47 \times 0 + 0.22 \times 0.3 = 0.38$$

$$[I_{Y(x_0)=2}]^{DK} = 0.31 \times 0 + 0.47 \times 0 + 0.22 \times 0.2 = 0.04$$

$$[I_{Y(x_0)=3}]^{DK} = 0.31 \times 0 + 0.47 \times 1 + 0.22 \times 0.1 = 0.49$$

$$[I_{Y(x_0)\geq 2}]^{DK} = 0.31 \times 0 + 0.47 \times 1 + 0.22 \times 0.3 = 0.54$$

$$[Y(x_0)]^{DK} = 0.31 \times 1 + 0.47 \times 3 + 0.22 \times 1.0 = 1.94.$$

So, according DK estimation, $Y(x_0)$ is estimated to 1.94, and the probability for the true $Y(x_0)$ to be equal: to 0 is 9%; to 1 is 38%; to 2 is 4%; to 3 is 49%.

4

Isofactorial model with no edge effects: orthogonal indicator residuals

In the model presented here, there are no edge effects: the edges of the set of points above a given cut-off are not poorer than the centre.

This is an isofactorial model, which means that DK (i.e. indicator cokriging) is obtained by kriging the factors, here the indicator residuals.

Isofactorial models

We have seen in Chapter 1 that cokriging different variables is simplified when the variables can be factorized. Then we need only to krige each factor separately.

The same is true for the indicator variables of a random function $Y(x)$. Isofactorial models are models where the different indicators are factorized. Then cokriging the indicators, i.e. DK, is equivalent to kriging the factors.

There are two types of isofactorial models: those with edge effects on the regions defined by cut-offs, and those without them.

Edge effects

Let us see what is meant by edge effects. As before, we consider a random function $Y(x)$ that takes four values: 0, 1, 2, 3. Let A_1 denote the set of points where the value of $Y(x)$ equals or exceeds 1. So for each point x in this set we have:

$$I_{x \in A_1} = I_{Y(x) \geq 1} = 1.$$

Of course, the value of $Y(x)$ for a point x inside A_1 might also equal or exceed 2. But this may have less chance of happening when the point x is near the edge of A_1. That is, there is an edge effect because the border areas of A_1 are, on average, of lower grade than is the centre. Fig. 4.1 illustrates this. In this case, the value at a point x inside A_1 depends on whether a neighbouring point $x + h$ belong to A_1 or not. In particular the probability:

$$P[Y(x) \geq 2 \,|\, Y(x) \geq 1,\ Y(x+h) < 1] = \frac{P[Y(x) \geq 2,\ Y(x+h) < 1]}{P[Y(x) \geq 1,\ Y(x+h) < 1]} \qquad (4.1)$$

will depend on h. In general, it increases with increasing h at least for short distances.

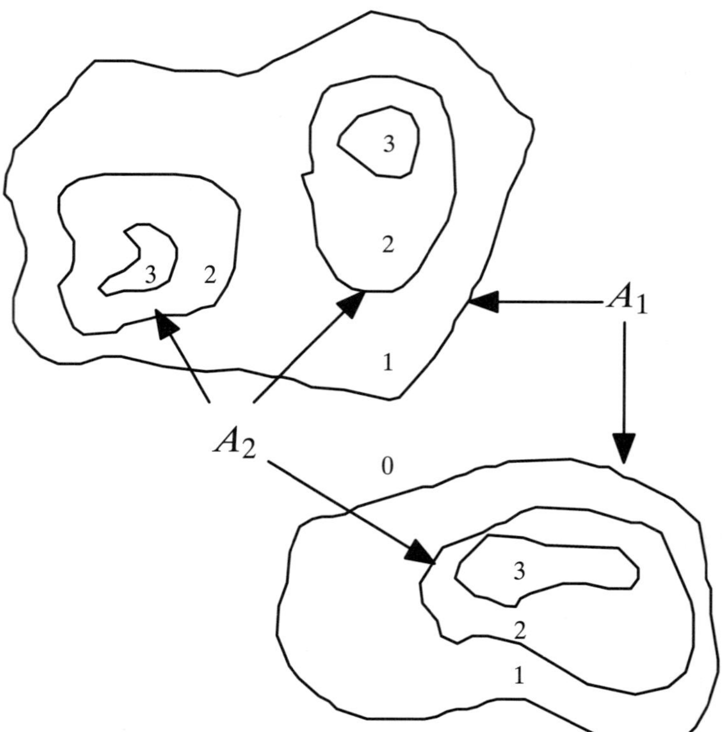

Fig. 4.1 Example of edge effects. In this case, the border areas of A_1 are poorer than the centre. The same is also true for A_2, the set consisting of points where $Y(x) \geq 2$. So there are edge effects on these sets.

This can be seen experimentally using the variograms of the indicators. For example, the variogram for the indicator $I(Y(x) \geq 1)$ can be written as:

$$\gamma_1(h) = \frac{1}{2}E[I_{Y(x)\geq 1} - I_{Y(x+h)\geq 1}]^2.$$

This is the mean value of a term which takes the value 1 if $Y(x) \geq 1$ and $Y(x + h) < 1$ or if $Y(x) < 1$ and $Y(x + h) \geq 1$, and takes the value 0 otherwise. So:

$$\gamma_1(h) = \frac{1}{2}\{P[Y(x) \geq 1,\ Y(x + h) < 1] + P[Y(x + h) \geq 1,\ Y(x) < 1]\}.$$

If we assume that the two probabilities are equal (that is, the distribution is symmetric in h):

$$\gamma_1(h) = P[Y(x) \geq 1,\ Y(x + h) < 1].$$

Similarly:

$$\gamma_{21}(h) = \frac{1}{2}E[I_{Y(x)\geq 2} - I_{Y(x+h)\geq 2}]\ [I_{Y(x)\geq 1} - I_{Y(x+h)\geq 1}]$$

is equal to:

$$P[Y(x) \geq 2,\ Y(x + h) < 1].$$

Consequently we have (for $h \neq 0$):

$$P[Y(x) \geq 2 \mid Y(x) \geq 1,\ Y(x + h) < 1] = \frac{\gamma_{21}(h)}{\gamma_1(h)}.$$

Using the indicator variograms it is easy to see whether there are edge effects. In general, the ratio given above increases with h for small distances, which shows the presence of an edge effect. As the border area of A_1 is, in general, poorer than its centre, going from a point where the value is small (e.g. 0) to one where it is larger generally means reaching an intermediate value (e.g. 1) rather than a large one (e.g. 3).

Absence of edge effect: orthogonal indicator residuals

We still consider a random function $Y(x)$ which takes the values 0, 1, 2, 3 and we assume that there are no edge effects on the sets A_1 (contains the points where $Y(x) \geq 1$) and A_2 (where $Y(x) \geq 2$).

For any point x in A_1 (i.e. where $Y(x) \geq 1$), the fact that its value equals or exceeds 2 is independent of a neighbouring point $x + h$ belonging to A_1. That is, the indicator $I(Y(x) \geq 2)$ is independent of $I(Y(x + h) \geq 1)$. So:

$$P[Y(x) \geq 2,\ Y(x + h) \geq 1 | Y(x) \geq 1]$$

$$= P[Y(x) \geq 2 | Y(x) \geq 1]\ \ P[Y(x + h) \geq 1 | Y(x) \geq 1].$$

And therefore:

$$\frac{P[Y(x) \geq 2,\ Y(x + h) \geq 1]}{P(Y(x) \geq 1)}$$

$$= \frac{P(Y(x) \geq 2)}{P(Y(x) \geq 1)}\ \frac{P[Y(x + h) \geq 1,\ Y(x) \geq 1]}{P(Y(x) \geq 1)}.$$

Using the notation introduced in Chapter 2, we have:

$$T^h_{21} = T^h_{11}\,\frac{T_2}{T_1}. \tag{4.2}$$

More generally, the bivariate distribution for $j>i$ of these models satisfies the relation:

$$T^h_{ji} = T^h_{ii}\,\frac{T_j}{T_i}. \tag{4.3}$$

As a general rule, the regression between indicator functions at a given point x can be written as:

$$E[I_{Y(x) \geq 2} | I_{Y(x) \geq 1}] = \frac{T_2}{T_1}\, I_{Y(x) \geq 1}.$$

This is clear as one of two things must happen:

(a) Either the conditioning variable $I(Y(x) \geq 1)$ is zero (i.e. $Y(x) < 1$) and so $I(Y(x) \geq 2) = 0$.

(b) Or alternatively $I(Y(x) \geq 1)$ is equal to 1, then $I(Y(x) \geq 2)$ is equal to 1 with the probability $P(Y(x) \geq 2 | Y(x) \geq 1) = T_2/T_1$ or to 0 otherwise, and so its mean value is T_2/T_1.

So the regression can be written as above. Its residual

$$I_{Y(x)\geq 2} - E[I_{Y(x)\geq 2}|I_{Y(x)\geq 1}] = I_{Y(x)\geq 2} - \frac{T_2}{T_1} I_{Y(x)\geq 1}$$

clearly has a zero mean and is uncorrelated with the conditioning variable $I(Y(x) \geq 1)$.

When there is no edge effect, this residual is also orthogonal to $I(Y(x + h) \geq 1)$:

$$E\left[\left(I_{Y(x)\geq 2} - \frac{T_2}{T_1} I_{Y(x)\geq 1}\right) I_{Y(x+h)\geq 1}\right] = T^h_{21} - \frac{T_2}{T_1} T^h_{11} = 0.$$

Such residuals play a special role in this model. Let us consider the family consisting of:

$$\begin{aligned} H_0(Y(x)) &= 1 \\ H_1(Y(x)) &= \frac{I_{Y(x)\geq 1}}{T_1} - 1 \\ H_2(Y(x)) &= \frac{I_{Y(x)\geq 2}}{T_2} - \frac{I_{Y(x)\geq 1}}{T_1} \\ H_3(Y(x)) &= \frac{I_{Y(x)\geq 3}}{T_3} - \frac{I_{Y(x)\geq 2}}{T_2}. \end{aligned}$$

For $i > 0$, $H_i(Y(x))$ is equal to the residual of $I(Y(x) \geq i)/T_i$ starting from the lower indicator $I(Y(x) \geq i - 1)$. The multiplicative factor $1/T_i$ permits the writings to be simplified. Then we have:

$$\sum_0^j H_i(Y(x)) = \frac{I_{Y(x)\geq j}}{T_j}$$

which means that these residuals form a generating family. Moreover, from (4.3), it follows that they are orthogonal. So these residuals factorize the indicators. They are the factors of the model whose bivariate distributions satisfy (4.3), which then is isofactorial. When using this model, we have only to krige each factor separately in order to cokrige the indicators or equivalently for the disjunctive kriging of any function:

$$\frac{[I_{Y(x)\geq j}]^{DK}}{T_j} = \sum_0^j [H_i(Y(x))]^K.$$

It is easy to extend these results to more than four values. However, if a variable takes many values, then these should be better grouped into a few classes.

The model can be tested experimentally by comparing the indicator cross-variograms $\gamma_{ij}(h)$ with the auto-variograms $\gamma_i(h)$ for $j > i$.

Numerical exercise

Let $Y(x)$ be a stationary random function taking the values 0, 1, 2, 3, with the probabilities $p_0 = 0.4$, $p_1 = 0.3$, $p_2 = 0.2$, $p_3 = 0.1$. Its mean is

$$E[Y(x)] = 0 \times 0.4 + 1 \times 0.3 + 2 \times 0.2 + 3 \times 0.1 = 1.0.$$

Knowing it is positive, $Y(x)$ takes the values 1, 2, 3 with probabilities 1/2, 1/3, 1/6, and its expectation is

$$E[Y(x) \mid Y(x) \geq 1] = 1 \times 1/2 + 2 \times 1/3 + 3 \times 1/6 = 10/6 = 1.7.$$

Suppose that the cross-variograms of its cumulated indicators $\gamma_{ij}(h)$ are proportional to the auto-variograms $\gamma_i(h)$ for $j > i$. So we are in a model where the indicator residuals are orthogonal. Once they have been calculated at the sampling points, we observe that:

(1) The first residual

$$H_1(Y(x)) = \frac{I_{Y(x) \geq 1}}{T_1} - 1,$$

and hence $I(Y(x) \geq 1)$, are spatially well structured. The covariance of $I(Y(x) \geq 1)$ is $0.24 \exp - |h|/12$.

(2) The other residuals

$$H_2(Y(x)) = \frac{I_{Y(x) \geq 2}}{T_2} - \frac{I_{Y(x) \geq 1}}{T_1}$$

and

$$H_3(Y(x)) = \frac{I_{Y(x) \geq 3}}{T_3} - \frac{I_{Y(x) \geq 2}}{T_2}$$

are pure nugget.

An example of this might be a set of mineralized veins, with a geometry which is structured, but within which the grades are completely random. The lowest cut-off, as it makes the distinction between the mineralized veins and their waste surroundings, defines the geometry of the veins. (This model could as well represent fish schools, or rainy areas.)

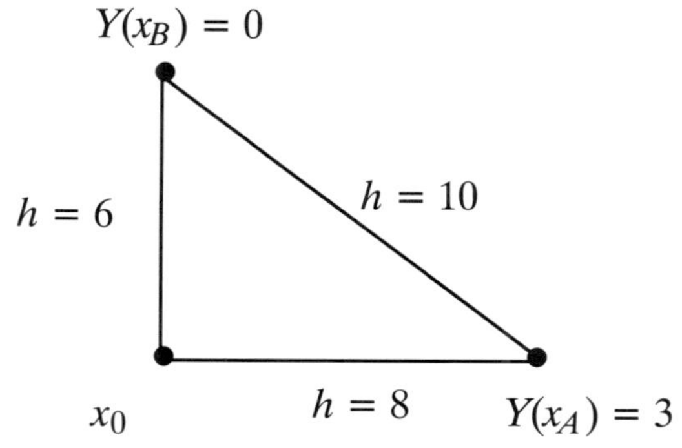

Fig. 4.2

We know the values of $Y(x)$ at the points x_A and x_B of Fig. 4.2 and we want to make estimations at point x_0 using DK.

(a) First residual, or indicator $I(Y(x) \geq 1)$ of the geometry of veins: its disjunctive kriging is equal to its kriging. Show that the kriging system is

$$\lambda_A\, 0.240 + \lambda_B\, 0.104 = 0.123$$
$$\lambda_A\, 0.104 + \lambda_B\, 0.240 = 0.146$$

and that the solution is $\lambda_A = 0.31$ and $\lambda_B = 0.47$.
Deduce the value of the DK estimate of $I(Y(x_0) \geq 1)$

$$[I_{Y(x_0)\geq 1}]^{DK} = [I_{Y(x_0)\geq 1}]^{K}$$
$$= 0.31 \times 1 + 0.47 \times 0 + 0.22 \times 0.6 = 0.44.$$

(b) Second residual:
at an unknown point, its kriged value is equal to 0. Deduce that the DK estimation of the indicator $I(Y(x_0) \geq 2)$ can be written

$$[I_{Y(x_0)\geq 2}]^{DK} = [I_{Y(x_0)\geq 1}]^{K}\, P(Y(x) \geq 2 \,|\, Y(x) \geq 1)$$
$$= 0.44\, \frac{0.3}{0.6} = 0.22.$$

(c) Third residual:
show that the DK estimation of the indicator $I(Y(x_0) \geq 3)$ can be written

$$[I_{Y(x_0)\geq 3}]^{DK} = [I_{Y(x_0)\geq 1}]^{K}\, P(Y(x) \geq 3 \,|\, Y(x) \geq 1)$$
$$= 0.44\, \frac{0.1}{0.6} = 0.07.$$

(d) Deduce that, according DK, there is a

0.56 probability for the true $Y(x_0)$ to equal 0
0.22 1
0.15 2
0.07 3.

Show that the DK of any function

$$f(Y(x_0)) = f_0\, I_{Y(x_0)=0} + f_1\, I_{Y(x_0)=1} + f_2\, I_{Y(x_0)=2} + f_3\, I_{Y(x_0)=3}$$

can be written

$$[f(Y(x_0))]^{DK} = 0.56\, f_0 + 0.22\, f_1 + 0.15\, f_2 + 0.07\, f_3.$$

In particular

$$[Y(x_0)]^{DK} = 0.56 \times 0 + 0.22 \times 1 + 0.15 \times 2 + 0.07 \times 3 = 0.73.$$

(e) According DK, there is a 44% probability for the true $Y(x_0)$ to be positive. Show that, in the case it would be positive, $Y(x_0)$ would take the values 1, 2, 3 with DK probabilities 0.22/0.44 = 1/2, 0.15/0.44 = 1/3, 0.07/0.44 = 1/6, which are just the a priori probabilities.

This agrees with the fact that the grade at a point x, knowing it belongs to the veins, is pure nugget, and then cannot be the subject of a really local estimation. Whether the positive neighbouring grades are 1, 2, or 3, has no influence on the estimation of the different indicators. The only really local estimation is the estimation of the geometry.

(f) Let us see in comparison what happens if we krige separately the indicators, for instance $I(Y(x_0) \geq 3)$. The covariance of this indicator being

$$\frac{1}{12}[\,1 \text{ nugget} + 0.08 \ \exp - |h|/12\,]$$

show that the kriging weights are $\lambda_A = 0.037$ and $\lambda_B = 0.044$ and deduce the kriged value

$$[I_{Y(x_0)\geq 3}]^K = 0.037 \times 1 + 0.044 \times 0 + 0.92 \times 0.1 = 0.13.$$

This kriging estimate, which does not take into account the other indicators, is much larger than the DK one (0.07).

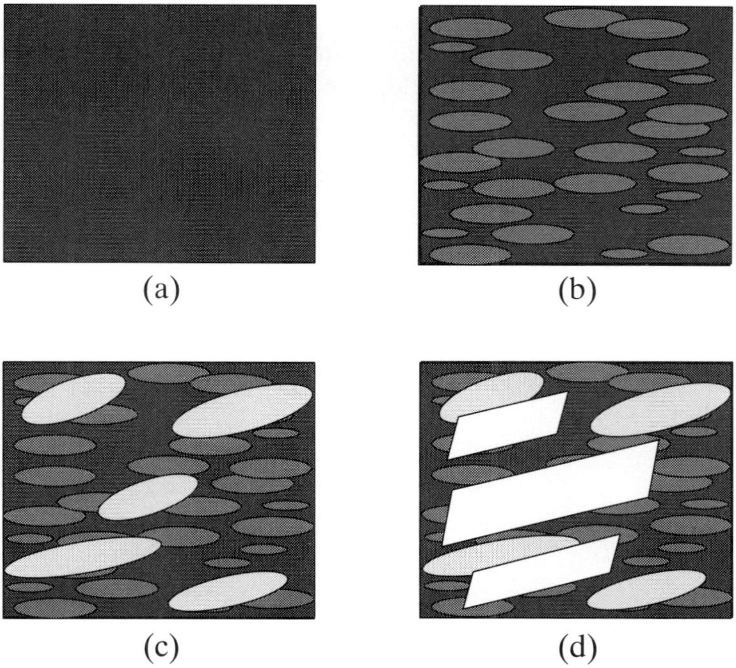

Fig. 4.3

Constructing the model

Constructing models with no edge effects on the sets A_i is simple.

(a) Start by assigning the largest value (e.g. 3) to all the points in the region.
(b) Then put the next value (e.g. 2) instead of 3 in some parts of the region.
(c) For another independent set of subregions, put the value 1 (instead of either 2 or 3).
(d) Finally, for a third set that is independent of both of the previous ones, put the smallest value (0).

These steps are shown in Fig. 4.3 (a) to (d).

The procedure can be reversed by starting from the smallest value and working to the largest. We obtain the same model, which is just inverted. In this case there are no edge effects for the sets of points *up to* each cut-off, and the factors are the residuals of each indicator $I(Y(x) < i)$ starting from the indicator for the *larger* cut-off grade $I(Y(x) < i + 1)$. The cross-variogram $\gamma_{ij}(h)$ between two indicators is now proportional to the variogram of the larger cut-off $\gamma_j(h)$ $(j > i)$.

The way Fig. 4.3 (d) has been constructed guarantees that there is no edge effect when leaving the small values. But when leaving the large values, we meet medium values preferentially to small ones. So there are edge effects in this direction. Note that in the mosaic model of Chapter 3 there are no edge effects in either direction (i.e. working upwards or downwards). Lastly, in the models described in the following chapters (and in Fig. 4.1 as the reader will realize), there are edge effects in both directions.

5

Isofactorial models with edge effects: the example of the gaussian model

The gaussian model is the most commonly used model with edge effects. It is an isofactorial model, the factors of which are the Hermite polynomials. DK is obtained by kriging these factors separately.

In practice the variable under study seldom has a normal distribution. So a transformation is needed to get to it.

Diffusion models

In general edge effects exist, and in order to go from a point with a value $(i-1)$ to a point with a value $(i+1)$, we go through points that take the intermediate value i. We then use 'diffusion models'. We shall not go into the theory in detail here, but it is based on stochastic diffusion processes. Once the process has reached a certain value i at a particular time, it can remain at that value, or move to a neighbouring state, either $(i+1)$ or $(i-1)$, but it cannot jump straight to any other value.

These models exist both when the values taken by the variable are discrete and when they are continuous. An example of the latter is the commonly used gaussian model.

Gaussian model

In this model the variable $Y(x)$ which is assumed to have zero mean and unit variance, has a normal (i.e. gaussian) distribution. That is, its p.d.f. is:

$$g(t) = \frac{1}{\sqrt{2\pi}} e^{-t^2/2}$$

which when graphed has the familiar form in Fig. 5.1. Its cumulative distribution function is the integral (Fig. 5.2):

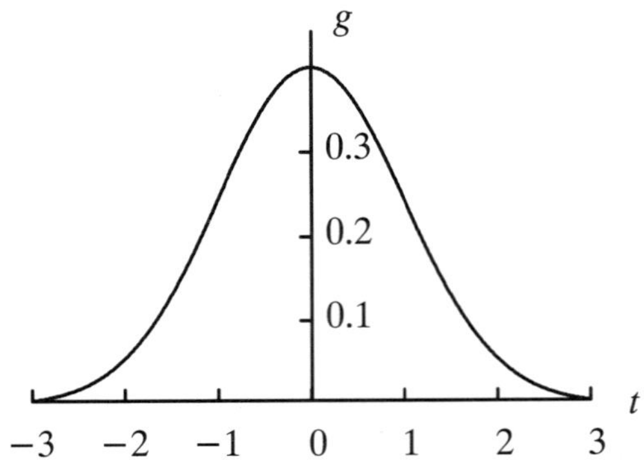

Fig. 5.1

$$G(u) = \int_{-\infty}^{u} g(t)\, dt$$

There is no convenient analytical expression for this function. However, its values have been worked out and tabulated, and it can be approximated numerically when using computers.

The pairs of points $(Y(x), Y(x + h))$ are assumed to have a bivariate normal distribution (i.e. any linear combination of them is normal) which is characterized by their correlation coefficient $\rho = \rho(h)$. Their p.d.f. can be written as:

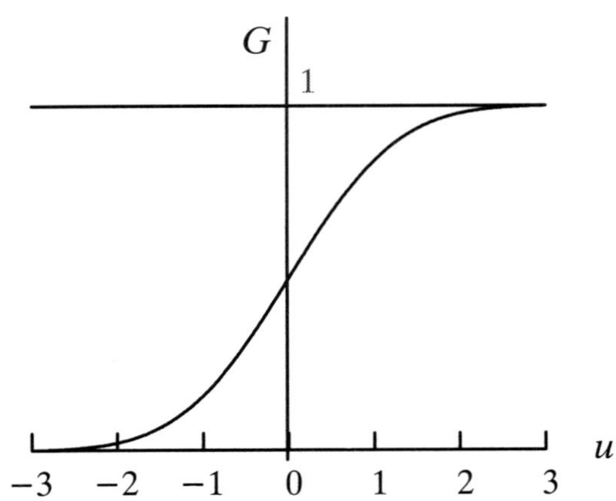

Fig. 5.2

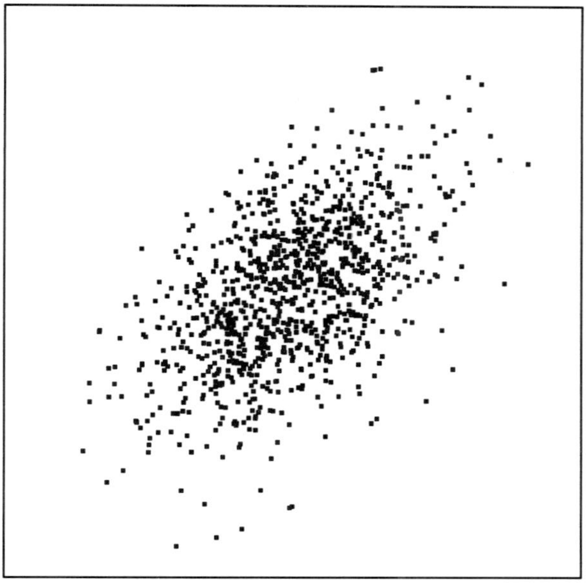

Fig. 5.3

$$g_\rho\ (t,u) = \frac{1}{2\pi\ \sqrt{1-\rho^2}}\ \exp\left(-\frac{t^2 + u^2 - 2tu\rho}{2(1-\rho^2)}\right).$$

Consequently their isodensity curves are the ellipses:

$$t^2 + u^2 - 2tu\rho = \text{constant}$$

The scatter diagram of the pairs of values is elliptic (Fig. 5.3), and all the narrower that the correlation is higher. When $\rho(h)$ is zero the variables $Y(x)$ and $Y(x+h)$ are independent and the scatter diagram becomes round.

Hermite polynomials

Hermite polynomials $H_n(Y(x))$ are polynomials that have special properties related to the normal distribution. They are defined by Rodrigues's formula ($n \geq 0$):

$$H_n(y) = \frac{1}{\sqrt{n!}\ g(y)}\ \frac{d^n\ g(y)}{dy^n} \tag{5.1}$$

where $\sqrt{n!}$ is a normalization factor. The Hermite polynomial $H_n(y)$ is a polynomial of degree n. More specifically,

$$H_0(y) = 1$$

$$H_1(y) = -y$$

$$H_2(y) = (y^2 - 1) / \sqrt{2}$$

The other polynomials are related and can be calculated using the recurrence relation ($n > 0$):

$$H_{n+1}(y) = -\frac{1}{\sqrt{n+1}}\, y\, H_n(y) - \sqrt{\frac{n}{n+1}}\, H_{n-1}(y)$$

So it is easy to calculate the Hermite polynomials starting from the value of the normal variable $Y(x)$. In practice, we generally need at most a few dozen of them.

Except for $H_0(Y(x))$ which is constant and equal to 1, their means are:

$$E[H_n(Y(x))] = \int H_n(y)\, g(y)\, dy = 0$$

and, because of the normalization factor, their variances are:

$$\text{var}\,[H_n(Y(x))] = E\{[H_n(Y(x))]^2\} = 1$$

Lastly when $p \neq n \geq 0$, the Hermite polynomials of a standard normal variable are orthogonal:

$$\text{cov}\,[H_p(Y(x)),\ H_n(Y(x))] = E[H_p(Y(x))\, H_n(Y(x))] = 0.$$

Expressing functions in terms of Hermite polynomials

Practically any function of $Y(x)$ can be expanded in terms of Hermite polynomials:

$$f[Y(x)] = f_0 + f_1\, H_1[Y(x)] + f_2\, H_2[Y(x)] + \ldots$$

$$= \sum_{n=0}^{\infty} f_n\, H_n[Y(x)].$$

Using the orthogonality of the Hermite polynomials, we find that

$$E[f(Y(x))\ H_n(Y(x))] = E\left\{[\sum_{p=0}^{\infty} f_p\ H_p(Y(x))]\ H_n(Y(x))\right\}$$

$$= \sum_{p=0}^{\infty} f_p\ E[H_p(Y(x))\ H_n(Y(x))]$$

is equal to f_n. This enables us to calculate these coefficients:

$$f_n = E[f(Y(x))\ H_n(Y(x))] = \int f(y)\ H_n(y)\ g(y)\ dy$$

for any given function f. In particular:

$$f_0 = E[f(Y(x))].$$

Exercise

Show that the variance of $f(Y(x))$ is given by:

$$\text{var}\ f[Y(x)] = \sum_{1}^{\infty} (f_n)^2.$$

Hint: develop

$$\text{var}\ f[Y(x)] = E[f[Y(x)] - f_0]^2 = E[\sum_{n=1}^{\infty} f_n\ H_n(Y(x))]^2$$

and use the orthogonality of the polynomials.

Example of the indicator function

For example the coefficient of order n of the indicator function $I(Y(x) < y_c)$ is:

$$f_n = \int I_{y < y_c}\ H_n(y)\ g(y)\ dy = \int_{-\infty}^{y_c} H_n(y)\ g(y)\ dy$$

We find for $n = 0$:

$$f_0 = G(y_c)$$

and for $n \geq 1$, from the definition (5.1) of the polynomials:

$$f_n = \frac{1}{\sqrt{n}} H_{n-1}(y_c)\, g(y_c).$$

Hence the expansion of the indicator:

$$I_{Y(x) < y_c} = G(y_c) + \sum_{n \geq 1} \frac{1}{\sqrt{n}} H_{n-1}(y_c)\, g(y_c)\, H_n[Y(x)]. \tag{5.2}$$

The expansion of the complementary indicator can be deduced

$$I_{Y(x) \geq y_c} = 1 - I_{Y(x) < y_c}$$

$$= 1 - G(y_c) - \sum_1 \frac{1}{\sqrt{n}} H_{n-1}(y_c)\, g(y_c)\, H_n[Y(x)]$$

with coefficients $1 - G(y_c)$ for $n = 0$, and $-\ (1/\sqrt{n})\ H_{n-1}(y_c)\ g(y_c)$ for $n \geq 1$.

So any function of $Y(x)$, in particular any indicator of $Y(x)$, can be expanded in terms of the generating family $H_n(Y(x))$. We then need only to cokrige $H_n(Y(x))$ in order to deduce the cokriging of the indicators and hence the disjunctive kriging of any function.

Structure of the Hermite polynomials

When the pairs $(Y(x), Y(x + h))$ are bivariate standard normal with correlation $\varrho(h)$, the Hermite polynomials have the following property:

$$E[H_n(Y(x + h)) \,|\, Y(x)] = [\rho(h)]^n\, H_n(Y(x)). \tag{5.3}$$

and so:

$$\begin{aligned} \text{cov}\ [H_p(Y(x)),\ H_n(Y(x + h))] &= E[H_p(Y(x))\ H_n(Y(x + h))] \\ &= E[H_p(Y(x))\ E(H_n(Y(x + h)) \,|\, Y(x))] \\ &= [\rho(h)]^n\ E[H_p(Y(x))\ H_n(Y(x))] \end{aligned}$$

Taking $p = n$ gives the formula

$$\text{cov}\ [H_n(Y(x)),\ H_n(Y(x + h))] = [\rho(h)]^n. \tag{5.4}$$

The spatial covariance of $H_n(Y(x))$ is equal to $[\rho(h)]^n$; that is, the covariance of $Y(x)$ raised to the nth power. As $\varrho(h) < \varrho(0) = 1$, the spatial dependence of $H_n(Y(x))$ decreases rapidly to nothing as n increases, i.e. the structure tends to pure nugget.

For $p \neq n$ we find

$$\mathrm{cov}\,[H_p(Y(x)), H_n(Y(x+h))] = E[H_p(Y(x))H_n(Y(x+h))] = 0 \quad (5.5)$$

So the Hermite polynomials for a bivariate standard normal pair also are orthogonal. There is no spatial correlation between polynomials of different orders.

Gaussian disjunctive kriging

As a consequence of this spatial non-correlation the Hermite polynomials are the factors of the (bi-)gaussian model, which then is isofactorial. Consequently the polynomials have only to be kriged separately to give the cokriging of the indicators and hence the DK of any function of $Y(x)$:

$$f[Y(x)]^{DK} = f_0 + f_1\,[H_1(Y(x))]^K + f_2\,[H_2(Y(x))]^K + \ldots \quad (5.6)$$

If we denote:

$x_\alpha,\ x_\beta \ldots$ the experimental points,
$\varrho_{\alpha\beta}$ the covariance $\varrho\,(x_\alpha - x_\beta)$ between $Y(x_\alpha)$ and $Y(x_\beta)$,
$H_n(Y_\alpha)$ the Hermite polynomials $H_n[Y(x_\alpha)]$,
we have:

$$[H_n(Y(x))]^K = \sum_\alpha \lambda_{n\alpha}\, H_n(Y_\alpha)$$

where the $\lambda_{n\alpha}$ satisfy the system:

$$\sum_\beta \lambda_{n\beta}\,\mathrm{cov}\,[H_n(Y_\alpha),\ H_n(Y_\beta)] = \mathrm{cov}\,[H_n(Y_\alpha),\ H_n(Y(x))] \qquad \equiv \alpha$$

i.e.

$$\sum_\beta \lambda_{n\beta}\,[\rho_{\alpha\beta}]^n = [\rho_{\alpha x}]^n \qquad \text{for all } \alpha$$

As the correlation structure $[\varrho(h)]^n$ of $H_n(Y(x))$ rapidly tends to one of pure nugget, the kriged estimator at an unknown point rapidly tends to its mean, that is, to zero. So even if the coefficients f_n are not negligible, we have to krige only fairly few polynomials in (5.6) to get the result. Usually it is less than a dozen.

Remark

It can be shown that the kriging variance of $H_n(Y(x))$ is:

$$\sigma^2_{Kn} = 1 - \sum_{\alpha} \lambda_{n\alpha} \, [\rho_{\alpha x}]^n$$

and that the DK estimation variance of $f[Y(x)]$ is given by:

$$\text{var} \; (f[Y(x)] - f[Y(x)]^{DK}) = \sum_{1}^{\infty} (f_n)^2 \, \sigma^2_{Kn}.$$

Numerical example

This example can be done as an exercise. Suppose the gaussian variable $Y(x)$ has the covariance $\varrho(h) = \exp -|h|/2$. We know the values of $Y(x)$ at the points x_A and x_B of Fig. 5.4 (a) and we want to estimate the unknown values of $Y(x)$ and $I(Y(x) < 0)$ at point x_0 using DK.

Kriging the first polynomial

The experimental values of $H_1(Y(x)) = -Y(x)$ and the correlations $\varrho(h)$ are given Fig. 5.4 (b). The weights λ_A and λ_B satisfy the system:

$$\lambda_A + \lambda_B \, 0.082 = 0.135$$

$$\lambda_A \, 0.082 + \lambda_B = 0.223.$$

We obtain $\lambda_A = 0.118$ and $\lambda_B = 0.213$ (and then a weighting factor of 67% for the zero mean of $H_1(Y(x))$). The kriged value of $H_1(Y(x_0))$ equals

$$[H_1(Y(x_0))]^K = 0.118\,(0) + 0.213\,(-1) = -0.213.$$

Kriging the second polynomial

The experimental values of $H_2(Y(x)) = (Y(x)^2 - 1)/\sqrt{2}$ and the correlations $[\rho(h)]^2$ are given Fig. 5.4 (c). We now obtain $\lambda_A = 0.018$ and $\lambda_B = 0.05$ and a weighting factor for the zero mean of 93%. We get

$$[H_2(Y(x_0))]^K = -0.013.$$

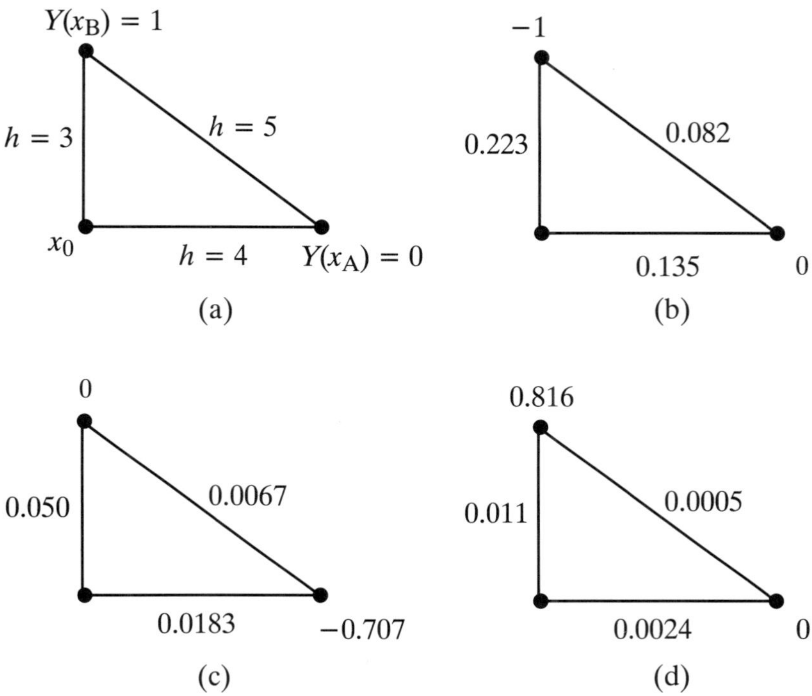

Fig. 5.4

Kriging the third polynomial

The experimental values of $H_3(Y(x))$ and the correlations $[\rho(h)]^3$ are given Fig. 5.4 (d). Hence $\lambda_A = 0.002$ and $\lambda_B = 0.011$ and a weight of the zero mean close to 99%. The kriged value is

$$[H_3(Y(x_0))]^K = 0.009.$$

Other polynomials

In this short example the correlations and then the kriging weights are practically zero beyond order 3. The kriged value of the corresponding polynomials will be taken as zero.

DK estimate of Y(x)

We just have now to replace the polynomials by their kriged values in the development of the functions we want to estimate. The development of $Y(x)$ is clearly reduced to $Y(x) = -H_1(Y(x))$ then

$$[Y(x_0)]^{DK} \;=\; -\;[H_1(Y(x_0))]^K = 0.21.$$

DK estimate of an indicator

The development of the indicator can be written (5.2):

$$I_{Y(x)<0} \;=\; G(0) + \sum_{n\geq 1} \frac{1}{\sqrt{n}}\; H_{n-1}(0)\; g(0)\; H_n[Y(x)]$$

that is:

$$0.5 \;+\; 0.4\; H_1(Y(x)) \;+\; 0\; H_2(Y(x)) \;-\; 0.163\; H_3(Y(x)) \;+\; \ldots$$

Its DK estimator $[I(Y(x_0)\;<\;0)]^{DK}$ can be deduced

$$0.5 + 0.4\;[H_1(Y(x_0))]^K + 0\;[H_2(Y(x_0))]^K - 0.163\;[H_3(Y(x_0))]^K + \;\ldots$$

i.e. $\quad 0.5 + 0.4\,(-\,0.213) + 0 - 0.163\,(0.009) + 0 = 0.413$

and also:

$$[I_{Y(x_0)\geq 0}]^{DK} = [1 - I_{Y(x_0)<0}]^{DK} = 1 - [I_{Y(x_0)<0}]^{DK} = 0.587.$$

To summarize DK gives an estimated value of 0.21 for $Y(x_0)$ and gives 41% of chance for the true unknown value of $Y(x_0)$ to be negative.

Gaussian anamorphosis or a transformation to normality

Gaussian disjunctive kriging presupposes that the variables ($Y(x)$, $Y(x+h)$) are bivariate normal, and consequently that $Y(x)$ is univariate normal. The variable studied, $Z(x)$, is very rarely normally distributed, so some sort of transformation (called anamorphosis) is required to convert it to a normal variable $Y(x)$. We then assume that the transformed pairs ($Y(x)$, $Y(x+h)$) are bivariate normal.

This transformation converts the histogram of data $Z(x)$ into a standard normal distribution, Fig. 5.5. On the cumulative distribution functions of $Z(x)$ and $Y(x)$, it associates each value z with the value y which corresponds to the same cumulative probability, Fig. 5.6. The function $z \;=\; \Phi(y)$ that relates z and y is called the anamorphosis function, Fig. 5.7.

Comments

(1) Knowing the anamorphosis of the variable $Z(x)$ is equivalent to knowing its distribution. In particular the anamorphosis records the irregularities that are present in the experimental distribution and create problems when one wants to represent the distribution with a given statistical law (normal, lognormal, gamma, . . .).
(2) The anamorphosis of a normal variable is linear (see formula (5.7)).
(3) The lognormal case is presented in Exercise 3.

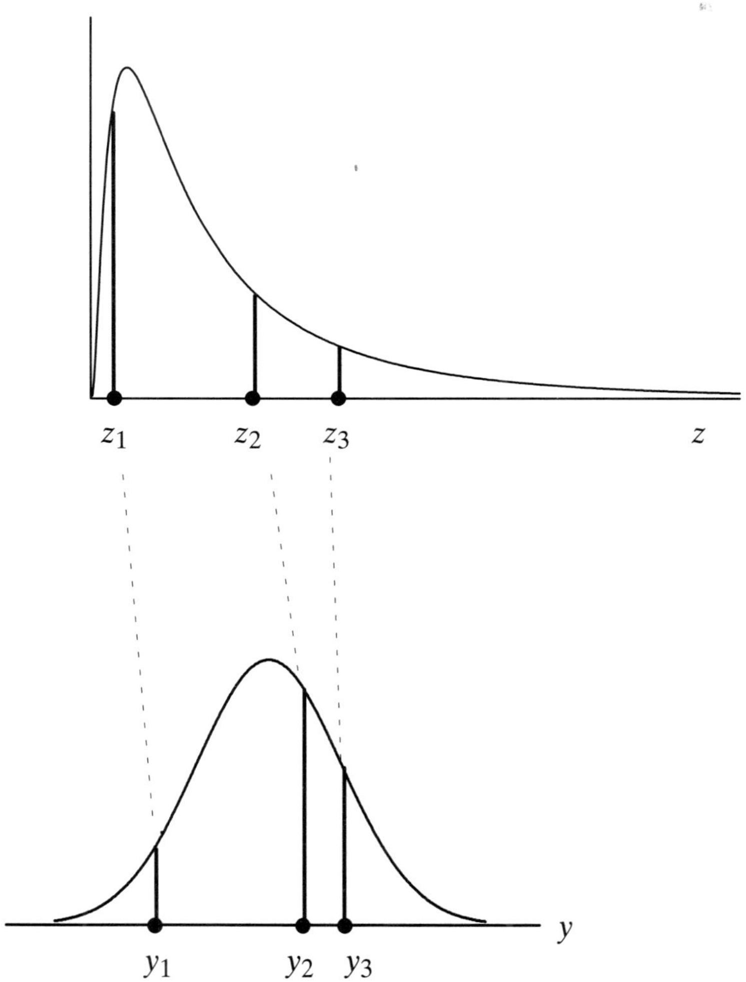

Fig. 5.5

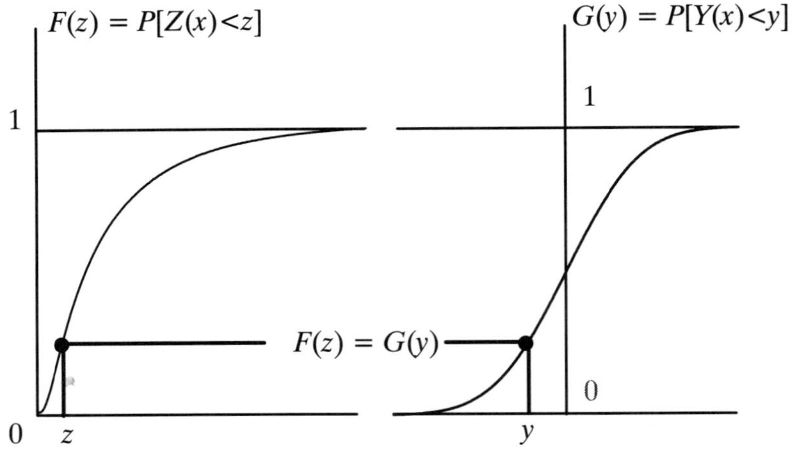

Fig. 5.6

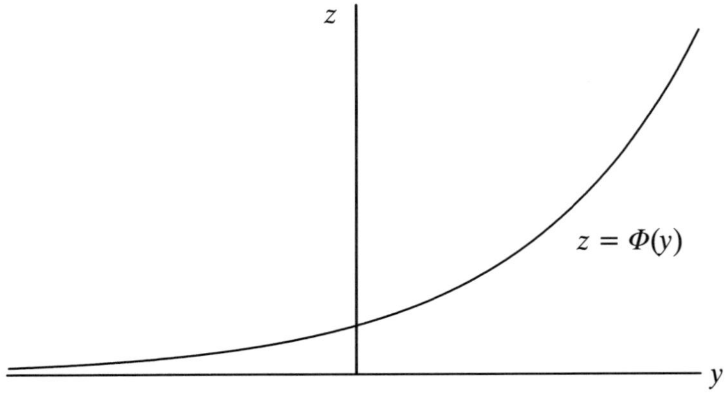

Fig. 5.7

Other diffusion models

It is not always possible to transform the variable $Z(x)$ to normality. Suppose for example that we are dealing with a concentration and that 50% of the values at the sampling points are zero. They correspond to the 50% of negative gaussian equivalents, but there is a problem deciding which gaussian value we should associate with each zero. An arbitrary choice for each gives a normal distribution, but the local estimates obtained from these values could well be arbitrary too. (Moreover, the bivariate distribution will not generally be normal.)

When the transformed variable $Y(x)$ is not bivariate normal, or when the distribution of $Z(x)$ contains 'atoms' (that is, classes of identical values), other isofactorial diffusion models should be used, for instance (Matheron 1973, 1976, 1984; Hu 1988):

The hermitian model which is a generalization of the preceding (bi-)gaussian one. The marginal distributions are normal and the factors are still Hermite polynomials.

The bivariate gamma, and more generally, the Laguerre models. Here the marginal distributions are gamma and the factors are the Laguerre polynomials.

The negative binomial models which are the discrete equivalent of the previous models.

Lastly a general method has been developed for constructing isofactorial models for the discrete case on an empirical basis. (Cf. Matheron 1984; Lajaunie and Lantuéjoul 1989).

Exercises

Here are some simple exercises designed to help readers work with the normal, bivariate normal and lognormal distributions.

(1) A non-standard normal variable

Suppose the mean and variance of a normally distributed random variable Z are m and σ^2 (instead of 0 and 1 for standard variables). In this case the variable $U = (Z - m)/\sigma$ is a standard normal variable. In other words, any normal variable can be written as:

$$Z = m + \sigma U \tag{5.7}$$

where m is the mean, σ^2 the variance, and U is the associated standard normal variable. This expression is very convenient as it allows us to express any normal variable in terms of a standard one for which we know the p.d.f. and the c.d.f.

Use this expression to show that the cumulative distribution function of Z can be written as:

$$P(Z < z) = G\left(\frac{z - m}{\sigma}\right).$$

Hence deduce the p.d.f. of Z by differentiating this:

$$\frac{1}{\sigma}\, g\left(\frac{z-m}{\sigma}\right) = \frac{1}{\sqrt{2\pi}\,\sigma} \exp -\frac{1}{2}\left(\frac{z-m}{\sigma}\right)^2.$$

(2) Conditioned normal variable (important!)

Let (Y_0, Y_1) be a pair of bivariate normal random variables with a standardized distribution and a correlation coefficient ϱ. Consider the random variable:

$$U = \frac{Y_0 - \rho Y_1}{\sqrt{1-\rho^2}}. \tag{5.8}$$

(a) Show that:

$E(U) = 0$

var $(U) = 1$

U is normally distributed

Consequently U is a standard normal variable.

(b) Secondly show that:

$$\text{cov}\,(U, Y_1) = 0.$$

So U and Y_1 are uncorrelated. Note that the pair (U, Y_1) is bivariate normal (any linear combination of U and Y_1 is normally distributed because it is also a linear combination of Y_0 and Y_1, which are bivariate normal). As U and Y_1 are uncorrelated and normally distributed, they are independent.

(c) Show that (5.8) can be written as:

$$Y_0 = \rho Y_1 + \sqrt{1-\rho^2}\; U$$

where U is a standard normal variable that is independent of Y_1. Consequently, if we know that $Y_1 = y_1$, we have:

$$(Y_0 | Y_1 = y_1) = \rho y_1 + \sqrt{1-\rho^2}\; U. \tag{5.9}$$

From this deduce that the variable Y_0 conditioned by $Y_1 = y_1$ is normally distributed with mean $E(Y_0|Y_1 = y_1) = \rho y_1$ and with variance $1 - \varrho^2$.

If $Y_1 = y_1$ then we can use the value $E(Y_0|Y_1 = y_1)$ as an estimate of Y_0. Note that the estimation variance of Y_0 given $Y_1 = y_1$, i.e. $1 - \varrho^2$ does not depend on y_1.

(d) Deduce the regression equation of Y_0 knowing Y_1, that is:

$$E(Y_0 | Y_1) = \rho Y_1$$

Note that it is linear.

(e) Use the operation (5.9) to find the expression for the estimator of $I(Y_0 < y)$ given $Y_1 = y_1$:

$$E(I_{Y_0<y} | Y_1 = y_1) = P(Y_0 < y | Y_1 = y_1) = G\left(\frac{y - \rho y_1}{\sqrt{1 - \rho^2}}\right)$$

(f) Also find the estimator of a function $f(Y_0)$ of Y_0:

$$E[f(Y_0) | Y_1 = y_1] = E[f(\rho y_1 + \sqrt{1 - \rho^2}\ U)]$$

$$= \int f(\rho y_1 + \sqrt{1 - \rho^2} u)\ g(u)\ du.$$

(3) Lognormal variable

A variable Z is lognormal if its natural logarithm $\log Z$ is normally distributed. Let μ and σ^2 be the mean and variance of $\log Z$. We can then write:

$$\log Z = \mu + \sigma Y$$

where Y is a standard normal variable. Consequently:

$$Z = e^{\mu + \sigma Y} = e^{\mu} e^{\sigma Y}.$$

So we see that in this particular case, the anamorphosis function $Z = \Phi(Y)$ is an exponential. In the subsequent calculations we shall often use:

$$E(e^{\lambda U}) = e^{\frac{\lambda^2}{2}} \tag{5.10}$$

if U is a standard normal variable. Proof:

$$E(e^{\lambda U}) = \int e^{\lambda u} g(u)\ du = \int e^{\lambda u} \frac{1}{\sqrt{2\pi}} e^{-u^2/2}\ du$$

$$= e^{\frac{\lambda^2}{2}} \int \frac{1}{\sqrt{2\pi}} e^{-(u-\lambda)^2/2}\ d(u - \lambda) = e^{\frac{\lambda^2}{2}} \int g(u - \lambda)\ d(u - \lambda) = e^{\frac{\lambda^2}{2}}.$$

(a) Show that the mean of Z can be written as:

$$M = E[Z] = e^{\mu + \sigma^2/2}.$$

From this deduce that Z can be written as:

$$Z = M \exp\left(\sigma Y - \frac{\sigma^2}{2}\right). \tag{5.11}$$

(b) Use a similar procedure to calculate $E(Z^2)$ and hence deduce the variance of Z:

$$\text{var } Z = M^2 \left(e^{\sigma^2} - 1\right).$$

Note that the logarithmic variance σ^2 depends only on the raw relative variance $(\text{var } Z)/M^2$.

(4) Bi-lognormal regression

Suppose that Z_0 and Z_1 are lognormal (5.11):

$$Z_0 = M_0 \exp\left(\sigma_0 Y_0 - \sigma_0^2/2\right)$$

$$Z_1 = M_1 \exp\left(\sigma_1 Y_1 - \sigma_1^2/2\right)$$

where Y_0 and Y_1 are bivariate standard normal with correlation ϱ. Knowing that $Z_1 = z_1$ is, of course, equivalent to knowing that $Y_1 = y_1$ with:

$$z_1 = M_1 \exp\left(\sigma_1 y_1 - \sigma_1^2/2\right).$$

From (5.9), show that the conditioned variable $(Z_0|Z_1 = z_1)$ can be written:

$$(Z_0|Z_1 = z_1) = \left[M_0 \exp\left(\sigma_0 Y_0 - \sigma_0^2/2\right)|Y_1 = y_1\right]$$

$$= M_0 \exp\left[\sigma_0\left(\rho y_1 + \sqrt{1-\rho^2}\,U\right) - \sigma_0^2/2\right] \tag{5.12}$$

where U is a standard normal variable.

(a) Use (5.10) to determine the mean of this distribution:

$$E(Z_0|Z_1 = z_1) = E(Z_0|Y_1 = y_1) = M_0 \exp\left(\sigma_0 \rho\, y_1 - \sigma_0^2 \rho^2/2\right)$$

(b) Given that $Z_1 = z_1$, we can use $\overset{*}{z_0} = E(Z_0|Z_1 = z_1)$ as an estimate of Z_0. Separate out $\overset{*}{z_0}$ as a factor in the formula (5.12):

$$(Z_0|Z_1 = z_1) = \overset{*}{z_0} \exp\Big(\sigma_0 \sqrt{1-\rho^2}\, U - \sigma_0^2 (1-\rho^2)/2\Big)$$

and show that this conditioned variable is also lognormal. Calculate the variance of this variable (see Exercise 3(b)):

$$(\overset{*}{z_0})^2 \left[\exp\big(\sigma_0^2 (1-\rho^2)\big) - 1\right].$$

This is also the estimation variance of Z_0 given that $Z_1 = z_1$. Note that in contrast to the normally distributed case this conditional estimation variance depends, through $\overset{*}{z_0}$, on the conditioning value z_1.

(c) Regression:

$$E(Z_0|Z_1) = E(Z_0|Y_1) = M_0 \exp\big(\sigma_0 \rho Y_1 - \sigma_0^2 \rho^2/2\big).$$

Note that it is not the initial form of Z_0, where Y_0 would have been replaced improperly by $E(Y_0|Y_1) = \rho Y_1$.
Show that the regression can be written as:

$$E(Z_0|Z_1) = c\,(Z_1)^{\frac{\rho\sigma_0}{\sigma_1}}$$

where c is a constant. Note that this regression is not linear.

Remark: the bilognormal regression is often used in uranium mining to model the relation between the radiometric measures and the chemical uranium grades.

6

More about the gaussian model

In the preceding chapter we saw that we had only to krige each of the Hermite polynomials $H_n(Y(x))$ separately in order to deduce the disjunctive kriging estimator of any function of $Y(x)$ provided that the pairs $(Y(x),\ Y(x+h))$ are bivariate normal. This chapter gives more details on disjunctive kriging that are needed for instance to program it.

Disjunctive kriging of the raw variable

The transformed variable $Y(x)$ obtained from the raw variable $Z(x)$ is assumed to be bivariate normal. As we saw earlier, disjunctive kriging allows us to estimate any function of $Y(x)$ by using its expansion in terms of Hermite polynomials. So we can estimate $Z(x) = \Phi[Y(x)]$ this way once we have its expansion:

$$Z(x) = \Phi[Y(x)] = \phi_0 + \phi_1 H_1[Y(x)] + \phi_2 H_2[Y(x)] + \dots$$

The function Φ can be determined graphically as was shown in Fig. 5.5 and 5.6. Here we shall express it in terms of the coefficients φ_n of its expansion. We now describe a way to calculate these (Lantuéjoul and Rivoirard 1984).

Suppose there are enough sample data for the distribution of $Z(x)$ to be known experimentally. Although there are often many different values of $Z(x)$, there are only a finite number of them. We can arrange them in increasing order:

$$z_1 < z_2 < z_3 < \dots < z_l$$

with the frequencies:

$$p_1 \quad p_2 \quad p_3 \quad \dots \quad p_l \qquad\qquad (\sum_i p_i = 1)$$

Their cumulative frequencies are:

$$F(z_1) = P(Z(x) < z_1) = 0$$

$$F(z_2) = P(Z(x) < z_2) = p_1$$

$$F(z_3) = P(Z(x) < z_3) = p_1 + p_2$$

$$\ldots$$

$$F(z_i) = P(Z(x) < z_i) = \sum_{1}^{i-1} p_j$$

$$\ldots$$

$$F(z_l) = P(Z(x) < z_l) = 1 - p_l.$$

These cumulative frequencies correspond to gaussian equivalents (normal scores) y_i with the same cumulative frequency:

$$F(z_i) = G(y_i).$$

So we have:

$$F(z_{i+1}) - F(z_i) = G(y_{i+1}) - G(y_i).$$

That is:

$$P[z_i \leq Z(x) < z_{i+1}] = P[y_i \leq Y(x) < y_{i+1}]$$

$$P[Z(x) = z_i] = P[y_i \leq Y(x) < y_{i+1}].$$

So $Z(x)$ will be equal to z_i when the gaussian equivalent lies between y_i and y_{i+1}. This allows us to determine the coefficients of the transformation:

$$\phi_0 = E[\Phi(Y(x))] = E[Z(x)] = \sum_i p_i \, z_i$$

$$\phi_n = E[Z(x) \, H_n(Y(x))] = \int \Phi(y) \, H_n(y) \, g(y) \, dy$$

$$= \sum_{i=1}^{l} \int_{y_i}^{y_{i+i}} z_i \, H_n(y) \, g(y) \, dy$$

$$= \sum_{i=1}^{l} z_i \left(\frac{1}{\sqrt{n}} H_{n-1}(y_{i+1}) \, g(y_{i+1}) - \frac{1}{\sqrt{n}} H_{n-1}(y_i) \, g(y_i) \right)$$

$$= \sum_{i=2}^{l} (z_{i-1} - z_i) \frac{1}{\sqrt{n}} H_{n-1}(y_i) \, g(y_i)$$

with $g(y_0) = g(-\infty) = 0$ and $g(y_{l+1}) = g(\infty) = 0$.

As we now know the coefficients, we can krige the $H_n(Y(x))$ to obtain the disjunctive kriging of $Z(x) = \Phi[Y(x)]$:

$$[Z(x)]^{DK} = \phi_0 + \phi_1 \, [H_1(Y(x))]^K + \phi_2 \, [H_2(Y(x))]^K + \ldots$$

Relation between the covariances

Let us consider the expansion of a function $f[Y(x)]$ in terms of Hermite polynomials:

$$f[Y(x)] = f_0 + \sum_{1}^{\infty} f_n \, H_n(Y(x))$$

with $f_0 = E[f(Y(x))]$. Its spatial covariance is:

$$\begin{aligned}\operatorname{cov}\left[f(Y(x)),\ f(Y(x+h))\right] &= E\left[(f[Y(x)] - f_0)(f[Y(x+h)] - f_0)\right] \\ &= E\left[\left(\sum_{n\geq 1} f_n \, H_n[Y(x)]\right)\left(\sum_{p\geq 1} f_p \, H_p[Y(x+h)]\right)\right] \\ &= \sum_{n\geq 1}\sum_{p\geq 1} f_n \, f_p \, E[H_n(Y(x)) \, H_p(Y(x+h))].\end{aligned}$$

Since the bivariate distribution $(Y(x),\ Y(x+h))$ is assumed to be normal the Hermite polynomials are orthogonal (5.5) and so we obtain:

$$\begin{aligned}\operatorname{cov}\left[f(Y(x)),\ f(Y(x+h))\right] &= \sum_{n\geq 1} (f_n)^2 \, E[H_n(Y(x)) \, H_n(Y(x+h))] \\ &= \sum_{n\geq 1} (f_n)^2 \, [\rho(h)]^n. \qquad (6.1)\end{aligned}$$

So there is a relationship between the covariance of the gaussian equivalent and that of any function of it. In particular, we get the following relation between the covariance of the gaussian equivalent obtained by anamorphosis, assumed to be bivariate normal, and that of the raw variable $Z(x) = \Phi[Y(x)]$:

$$\operatorname{cov}[Z(x),\ Z(x+h)] = \sum_{n\geq 1} (\phi_n)^2 \, [\rho(h)]^n \qquad (6.2)$$

This relation can even be used to test the binormality of the pairs $(Y(x),\ Y(x+h))$.

Remark

As n increases, $[\rho(h)]^n$ becomes spatially less dependent and tends to pure nugget. In particular, for $n > 1$, $[\rho(h)]^n$ is less structured than $\rho(h)$, which is the covariance of the gaussian variable. So $Z(x)$, from (6.2), or more generally any function of $Y(x)$, from (6.1), proves to be less structured than the gaussian variable.

Indicators

In many problems involving thresholds, we want to estimate whether the unknown value $Z(x)$ of the variable at point x is more or less than a certain cut-off z_c. In other words, we want to estimate the indicator $I(Z(x) < z_c)$. Note again that this is not a question of knowing if an estimated value $Z(x)^*$ is less than the cut-off.

In the gaussian model, the situation $Y(x) < y_c$ is equivalent to $Z(x) = \Phi[Y(x)] < \Phi(y_c)$. By choosing y_c so that $z_c = \Phi(y_c)$, that is, the cumulative frequency $G(y_c)$ equals $F(z_c)$, we have:

$$I_{Z(x) < z_c} = I_{Y(x) < y_c}$$

We know the Hermite polynomial expansion of this (5.2):

$$G(y_c) + \sum_{1}^{\infty} \frac{1}{\sqrt{n}} H_{n-1}(y_c)\, g(y_c)\, H_n[Y(x)]$$

So the DK estimator is obtained by kriging each of the $H_n(Y(x))$ separately (5.6):

$$[I_{Z(x) < z_c}]^{KD} = G(y_c) + \sum_{1} \frac{1}{\sqrt{n}} H_{n-1}(y_c)\, g(y_c)\, [H_n(Y(x))]^K$$

As the kriged values tend to 0 rapidly as n increases, only a few terms are required, even though the coefficients $(1/\sqrt{n})\, H_{n-1}(y_c)\, g(y_c)$ are not negligible.

Ore and metal quantities

Let $Z(x)$ represent the grade of an ore at a point x. As we have seen in (2.2), the quantities of ore and metal recovered above a cut-off z_c are, apart from a multiplicative constant, equal to:

$$I_{Z(x) \geq z_c}$$

$$Z(x)\ I_{Z(x) \geq z_c}$$

If we let y_c be the gaussian equivalent of z_c, and $Y(x)$ be the gaussian transform of $Z(x)$, then the quantity of ore becomes:

$$I_{Y(x) \geq y_c}.$$

That is (5.2):

$$1 - I_{Y(x) < y_c} = 1 - G(y_c) - \sum_{1} \frac{1}{\sqrt{n}} H_{n-1}(y_c)\ g(y_c)\ H_n[Y(x)].$$

Similarly the quantity of metal can be written as:

$$Z(x)\ I_{Z(x) \geq z_c} = \Phi[Y(x)]\ I_{Y(x) \geq y_c} = \left[\sum_{p \geq 0} \phi_p\ H_p[Y(x)]\right] I_{Y(x) \geq y_c}.$$

This is a function of $Y(x)$. It can be expanded in terms of Hermite polynomials:

$$Z(x)\ I_{Z(x) \geq z_c} = \sum_{n \geq 0} q_n\ H_n[Y(x)].$$

The coefficients q_n, depending on y_c, can be written as:

$$\begin{aligned} q_n &= E[\Phi(Y(x))\ I_{Y(x) \geq y_c}\ H_n(Y(x))] \\ &= \int \left[\sum_{p \geq 0} \phi_p\ H_p(y)\right] I_{y \geq y_c}\ H_n(y)\ g(y)\ dy \\ &= \sum_{p \geq 0} \phi_p \int_{y_c} H_p(y)\ H_n(y)\ g(y)\ dy. \end{aligned}$$

If we let $U_p^n(y_c)$ denote the integral in the formula above, then:

$$q_n = \sum_{p \geq 0} \phi_p\ U_p^n(y_c).$$

Using the relation:

$$U_p^n = \frac{-1}{\sqrt{n}} H_p\ H_{n-1}\ g + \sqrt{\frac{p}{n}}\ U_{p-1}^{n-1}$$

and the fact that $U_p^n = U_n^p$ we can get a recurrence formula from $U^0_{|n-p|}(y_c)$:

$$U_0^0(y_c) = 1 - G(y_c)$$

$$U_k^0(y_c) = \frac{-1}{\sqrt{k}} H_{k-1}(y_c)\ g(y_c)$$

Once we know the expansions of the quantity of ore and of metal, we get their DK estimates by kriging the $H_n(Y(x))$ separately:

$$[I_{Z(x)\geq z_c}]^{DK} = 1 - G(y_c) - \sum_1 \frac{1}{\sqrt{n}} H_{n-1}(y_c)\ g(y_c)\ [H_n(Y(x))]^K$$

$$[Z(x)\ I_{Z(x)\geq z_c}]^{DK} = q_0 + \sum_1 q_n\ [H_n(Y(x))]^K$$

7

DK and conditional expectation

In Chapter 2 we have seen how DK makes it possible to estimate functions $f[Y(x)]$ of a variable $Y(x)$. Here we will see another estimator, more consistent and theoretically more powerful, the conditional expectation. However we are able to use this estimator only in the case of multivariate normal distributions.

Consistencies and inconsistencies of DK

Disjunctive kriging is just the same as cokriging indicators. As we have seen in Chapter 2 this ensures a certain amount of consistency between the estimates of the different functions. In particular we have:

$$(I_{y_1 \leq Y(x) < y_2})^{DK} = (I_{Y(x) < y_2})^{DK} - (I_{Y(x) < y_1})^{DK}.$$

But kriging a positive-valued variable does not always give positive kriged values, nor does cokriging indicators necessarily give results that lie between 0 and 1. Sometimes the DK estimate of $I(Y(x) < y)$ for small values of y turns out to be negative, and similarly the estimate is sometimes greater than 1 for large values of y. This is clearly a nuisance.

The conditional expectation in a multinormal model

Suppose that we want to estimate the indicator $I(y_1 \leq Y(x) < y_2)$ using the values $Y(x_a) = y_\alpha$ which are available around the point x. Theoretically the best estimator is its expectation conditionally to the data:

$$[I_{y_1 \leq Y(x) < y_2}]^{CE} = E[I_{y_1 \leq Y(x) < y_2} | Y(x_a) = y_a \ldots]$$

$$= P[y_1 \leq Y(x) < y_2 | Y(x_a) = y_a \ldots].$$

That is, it is the conditional probability of having $y_1 \leq Y(x) < y_2$ given the $Y(x_a) = y_\alpha \ldots$ But this conditional distribution can be determined only for a multinormal random function.

In that case, the multivariate distribution of variables like ($Y(x)$, $Y(x_\alpha)$, . . .) is multinormal; i.e. any linear combination of these variables is also normally distributed. Consequently, the conditional distribution of $Y(x)$ given the $Y(x_\alpha) = y_\alpha$ . . . is normal, with a mean value equal to its kriged estimate:

$$y(x)^K = \sum \lambda^\alpha y_\alpha$$

and with its variance equal to the kriging variance σ_K^2. The reader will notice that this conditional multinormal distribution generalizes the conditional binormal one which has been studied in Exercise 2, Chapter 5. Its p.d.f. can be written as:

$$g(y \mid y_\alpha \ldots) = \frac{1}{\sqrt{2\pi}\ \sigma_K} \exp - \frac{1}{2}\left[\frac{y - \sum \lambda^\alpha y_\alpha}{\sigma_K}\right]^2$$

and the estimate of $I(y_1 \leq Y(x) < y_2)$ is then equal to:

$$\int_{y_1}^{y_2} g(y \mid y_\alpha \ldots)\, dy.$$

Now this is a probability and so its value lies between 0 and 1. If we put $u = (y - \Sigma\, \lambda^\alpha y_\alpha)/\sigma_K$, we can easily show that this is:

$$G\left[\frac{y_2 - \sum \lambda^\alpha y_\alpha}{\sigma_K}\right] - G\left[\frac{y_1 - \sum \lambda^\alpha y_\alpha}{\sigma_K}\right].$$

Similarly the conditional expectation of any function $f[Y(x)]$ can be written as:

$$\begin{aligned}[\, f[Y(x)]]^{EC} &= E[f[Y(x)] \mid Y(x_\alpha) = y_\alpha \ldots] \\ &= \int f(y)\, g(y \mid y_\alpha \ldots)\, dy \\ &= \int f(\textstyle\sum \lambda^\alpha y_\alpha + \sigma_K u)\, g(u)\, du. \qquad (7.1)\end{aligned}$$

So if $f[Y(x)]$ is a positive function this estimate will be positive. Similarly, if we are estimating two functions with one always greater than the other the difference between the two will always be positive.

So this estimator (which is the conditional expectation) will always ensure that the results are consistent, and as such, it is preferable to the bivariate normal DK. However the multivariate hypothesis that is theoretically required is much more restrictive than the one on the bivariate distributions.

Moreover DK is valuable for two reasons:

(1) It allows us to make estimations outside the normal framework (when it is no longer possible to obtain the multivariate conditional expectation).

(2) As we shall see in the next chapter, it is easy to estimate regularized variables using DK.

Numerical example of CE

Let us consider the same example as for DK (Chapter 5, Fig. 5.4(a)). The kriged value of $Y(x_0)$ from $Y(x_A) = 0$ and $Y(x_B) = 1$ is already known (apart from the sign, it is the same as for $H_1(Y(x_0)) = -Y(x_0)$):

$$[Y(x_0)]^K = 0.21$$

and its kriging variance is:

$$\begin{aligned}\sigma_K^2 &= 1 - \lambda_A \rho_{OA} - \lambda_B \rho_{OB}\\ &= 1 - 0.118(0.135) - 0.213(0.223) = 0.937.\end{aligned}$$

Under multigaussian hypothesis and conditional on $Y(x_A) = 0$ and $Y(x_B) = 1$, $Y(x_0)$ is gaussian with mean $[Y(x_0)]^K = 0.21$ and variance $\sigma_K^2 = 0.937$. So it can be written

$$Y(x_0) = 0.21 + \sqrt{0.937}\,U$$

with U standard gaussian. We then obtain the conditional expectation estimates:

$$\begin{aligned}[Y(x_0)]^{EC} &= [Y(x_0)]^K = 0.21\\ [I_{Y(x_0)<0}]^{EC} &= E[I_{Y(x_0)<0} \mid Y(x_A) = 0,\ Y(x_B) = 1]\\ &= P[Y(x_0) < 0 \mid Y(x_A) = 0,\ Y(x_B) = 1]\\ &= P[0.21 + \sqrt{0.937}\,U < 0]\\ &= P[U < -0.22] = \mathrm{G}(-0.22) = 0.41.\end{aligned}$$

8

Estimation of regularized variables

In linear geostatistics it is possible to krige a regularized variable without kriging each of its points. This is also true for DK. The regularized value of a function $f[Y(x)]$ can be estimated directly without having to estimate each point.

Kriging

In order to krige a regularized variable:

$$Z(V) = \frac{1}{V}\int_V Z(x)\,dx$$

we do not have to krige individually each of the points inside V even though it is equivalent. We can krige directly the regularized variable. We just replace the kriging system for each point x :

$$\sum_\beta \lambda_\beta \operatorname{cov}(Z_\alpha, Z_\beta) = \operatorname{cov}(Z_\alpha, Z(x)) \qquad \text{for all } \alpha$$

by the system for the block V:

$$\sum_\beta \lambda_\beta \operatorname{cov}(Z_\alpha, Z_\beta) = \operatorname{cov}(Z_\alpha, Z(V)) \qquad \text{for all } \alpha$$

where the right-hand side member is:

$$\operatorname{cov}(Z_\alpha, Z(V)) = \frac{1}{V}\int_V \operatorname{cov}(Z_\alpha, Z(x))\,dx.$$

Disjunctive kriging

This property is very useful for estimating regularized variables using DK. This may be done in a gaussian framework, as it is here, or in a different model. Suppose that we want to estimate the regularized variable:

$$\frac{1}{V}\int_V f(Y(x))\ dx.$$

Since:

$$f[Y(x)] = f_0 + \sum_1 f_n\ H_n[Y(x)]$$

we can deduce that:

$$\frac{1}{V}\int_V f(Y(x))\ dx = f_0 + \sum_1 f_n \left(\frac{1}{V}\int_V H_n(Y(x))\ dx\right).$$

To obtain the DK estimate we only need to krige each of the terms independently:

$$\frac{1}{V}\int_V H_n(Y(x))\ dx.$$

So:

$$\left(\frac{1}{V}\int_V f(Y(x))\ dx\right)^{DK} = f_0 + \sum_1 f_n \left(\frac{1}{V}\int_V H_n(Y(x))\ dx\right)^{K}.$$

The result of this would be exactly the same as the average of the DK estimates of $f[Y(x)]$ at each point.

Example: DK of a proportion

Suppose we want to estimate the regularized indicator:

$$\frac{1}{V}\int_V I_{Y(x)<y}\ dx.$$

This is the average of all the indicators (which take the values 0 or 1) inside V. As such it represents the proportion of points where $Y(x)$ is less than y.

From (5.2) we can see that:

$$\frac{1}{V}\int_V I_{Y(x)<y}\ dx = G(y) + \sum_1 \frac{H_{n-1}(y)\ g(y)}{\sqrt{n}} \left(\frac{1}{V}\int_V H_n(Y(x))\ dx\right)$$

The kriged estimate of $\frac{1}{V}\int_V H_n(Y(x))\ dx$ comes from the system:

$$\sum_\beta \lambda_{n\beta} \operatorname{cov}\ [H_n(Y_\alpha),\ H_n(Y_\beta)] = \operatorname{cov}\left[H_n(Y_\alpha),\ \frac{1}{V}\int_V H_n(Y(x))\ dx\right]$$

i.e.:

$$\sum_\beta \lambda_{n\beta} \operatorname{cov}\ [H_n(Y_\alpha),\ H_n(Y_\beta)] = \frac{1}{V}\int_V \operatorname{cov}\ [H_n(Y_\alpha),\ H_n(Y(x))]\ dx.$$

Since the covariance of $H_n(Y(x))$ is $[\rho(h)]^n$, this becomes:

$$\sum_\beta \lambda_{n\beta}\ [\rho_{\alpha\beta}]^n = \frac{1}{V}\int_V [\rho_{\alpha x}]^n\ dx.$$

Hence the DK estimator of the proportion is:

$$\left(\frac{1}{V}\int I_{Y(x)<y}\ dx\right)^{DK}$$

$$= G(y) + \sum_1 \frac{1}{\sqrt{n}} H_{n-1}(y)\ g(y) \left(\frac{1}{V}\int_V H_n(Y(x))\ dx\right)^K$$

$$= G(y) + \sum_1 \frac{1}{\sqrt{n}} H_{n-1}(y)\ g(y) \left(\sum_\alpha \lambda_{n\alpha}\ H_n(Y_\alpha)\right).$$

In practice, we need to krige only a few factors, that is, to solve only a few systems.

In comparison, if we want to estimate the same proportion via the conditional expectation we must estimate the variable $I(Y(x) < y)$ at a large enough number of points for it to be representative of the integral over V. So we have to solve the kriging system for each of these points (but the matrix on the left is the same and only has to be inverted once).

Important comment

In the gaussian anamorphosed model we can now estimate the indicator

$$I_{Z(x)<z} = I_{Y(x)<y}$$

(where $z = \Phi(y)$) either using DK or conditional expectation. Similarly we can also estimate the proportion of points in V where $Z(x) < z$:

$$\frac{1}{V}\int_V I_{Z(x)<z}\,dx = \frac{1}{V}\int_V I_{Y(x)<y}\,dx.$$

However we cannot estimate indicators on a regularized support such as:

$$I_{Z(V)<z}$$

For this, a change of support model is required (Chapters 10 to 12).

9

The problems of stationarity in point estimation

In linear geostatistics the condition $\Sigma\, \lambda_\alpha = 1$ is normally used when kriging a variable. This makes the local estimation more robust against a lack of stationarity.

Using such conditions in non-linear point estimation requires caution. Only in the mosaic model are there no problems. In DK estimation for example, these conditions lead to problems of convergence that must be looked at closely. In the case of the conditional expectation, the condition $\Sigma\, \lambda_\alpha = 1$ is justified only when estimating a lognormal variable (lognormal kriging).

Kriging

Let us recapitulate briefly on linear kriging. For simple kriging (i.e. kriging with a known mean) we have:

$$Z(x)^{SK} - E[Z(x)] = \sum_\alpha \lambda_\alpha\, [Z_\alpha - E(Z_\alpha)].$$

As the expectations $E[Z(x)]$ and $E[Z_\alpha]$ are known and are equal to each other (= m), we have:

$$Z(x)^{SK} = \sum_\alpha \lambda_\alpha\, Z_\alpha + \left(1 - \sum \lambda_\alpha\right) m.$$

This shows that SK gives a weighting factor $\lambda_m = 1 - \Sigma\, \lambda_\alpha$ to the known mean m. When the samples are fairly close (relative to the variogram structure) this weight is small. It increases as the information becomes more sparse, and the mean tends to counterbalance the lack of information: its role is more important in areas where there is less information. This is the consequence of the stationarity hypothesis which gives a local meaning to the mean m. In mining this causes problems in areas that are under-sampled because they are poorer than average and hence less economically profitable.

(This should make us reconsider the stationarity hypothesis.) Even in cases where the sampling is regularly spaced, the hypothesis of stationarity which attributes a local significance to the mean value that has been calculated over the whole deposit, may seem too strong.

We then use kriging with an unknown mean. (This is called ordinary kriging.) In this case the sum of the kriging weights of the neighbouring data is 100%:

$$\sum_{\alpha} \lambda_{\alpha} = 1.$$

The kriging estimator is $Z(x)^{OK} = \Sigma\, \lambda_{\alpha}\, Z_{\alpha}$. It is unbiased whatever the value of the unknown mean $m = E[Z(x)] = E[Z_{\alpha}]$. So:

$$E[Z(x) - Z(x)^{OK}] = E[Z(x) - \sum \lambda_{\alpha}\, Z_{\alpha}] = 0.$$

In other words, we assume that the mean can take any value, and that this may vary from one area to another but must be (approximately) constant over areas the size of the kriging neighbourhood. This excludes systematic trend or drift at this scale (working with drift is the purpose of the non-stationary part of geostatistics).

Cokriging

What is true for kriging single variables is also true for cokriging. For example, if the means are known, we have:

$$[Z_1(x) - E(Z_1(x))]^{CK} = \sum_{\alpha} \lambda_{1\alpha}\, [Z_{1\alpha} - E(Z_{1\alpha})] + \sum_{\beta} \lambda_{2\beta}\, [Z_{2\beta} - E(Z_{2\beta})].$$

So if $E(Z_1(x)) = E(Z_{1\alpha}) = m_1$ and similarly $E(Z_{2\beta}) = m_2$, then:

$$Z_1(x)^{CK} = \sum_{\alpha} \lambda_{1\alpha}\, Z_{1\alpha} + \sum_{\beta} \lambda_{2\beta}\, Z_{2\beta} + (1 - \sum_{\alpha} \lambda_{1\alpha})\, m_1 - (\sum_{\beta} \lambda_{2\beta})\, m_2.$$

Here α describes the data on the variable Z_1, and β those on Z_2, which need not to be located at the same points.

If instead we assume that the means are not known then we use the non-bias conditions:

$$\sum_{\alpha} \lambda_{1\alpha} = 1 \qquad \sum_{\beta} \lambda_{2\beta} = 0$$

Note that the condition $\Sigma_\alpha \lambda_{1\alpha} = 1$ requires at least one value of Z_1 inside the kriging neighbourhood.

Disjunctive kriging

Let us now consider the cases of indicators and of disjunctive kriging. In the mosaic model imposing the condition $\Sigma\, \lambda_\alpha = 1$ on the single system guarantees the unbiasedness of the estimator for all the indicators or for any function of the variable such as $f[Y(x)]$:

$$E[f(Y(x))\,^*] = \sum \lambda_\alpha\, E[f(Y_\alpha)] = E[f(Y(x))]$$

since the means, $E[f(Y(x))]$ and $E[f(Y_\alpha)]$, are locally the same.

In the isofactorial models any indicator or any function of the variable can be expressed in terms of the factors:

$$f[Y(x)] = a_0 + \sum_{1} a_n\, H_n(Y(x)).$$

When we impose the condition

$$\sum_{\alpha} \lambda_{n\alpha} = 1$$

on the kriging system of each of the factors:

$$[H_n(Y(x))]^K = \sum_{\alpha} \lambda_{n\alpha}\, H_n(Y_\alpha)$$

we have:

$$E[H_n(Y(x))]^K = \sum_{\alpha} \lambda_{n\alpha}\, E[H_n(Y_\alpha)] = E[H_n(Y(x))]$$

and then:

$$\begin{aligned} E[f(Y(x))\,^*] &= a_0 + \sum_{1} a_n\, E[H_n(Y(x))]^K \\ &= a_0 + \sum_{1} a_n\, E[H_n(Y(x))] \\ &= E[f(Y(x))]. \end{aligned}$$

So the estimator is unbiased whatever the local means of the factors. If, however, the means of the factors are (locally) unknown then so are those of the indicators and so is the (local) distribution of the variable. Taking gaussian DK as an example, this means that the 'gaussian' equivalent $Y(x)$ does not necessarily have (locally) a mean of zero and a unit variance, and does not even have (locally) a normal distribution. The factors (here the Hermite polynomials) are not necessarily orthogonal, which means that the optimality of the estimator is no longer guaranteed.

Moreover the unbiasedness conditions for gaussian DK have to be used carefully. Any function $f[Y(x)]$ can be expressed in terms of (an infinite number of) Hermite polynomials:

$$f[Y(x)] = f_0 + \sum_{n=1}^{\infty} f_n \, H_n[Y(x)]$$

As the structure of the Hermite polynomials tends rapidly to pure nugget with increasing n, kriging $H_n[Y(x)]$ at any unknown point *without* any unbiasedness condition soon returns to the mean value (i.e. zero) say when $n > L$. This is why so few polynomials (L) actually need be kriged.

The situation is different when an unbiasedness condition is introduced into the kriging system. In this case, for $n > L$, we have:

$$[H_n(Y(x))]^K = \frac{1}{N} \sum_{\alpha} H_n(Y_\alpha)$$

where N is the number of sampling points x_α in the kriging neighbourhood, and there is no reason for these quantities to be zero. To turn the difficulty of having an infinity of terms, we can use the relation:

$$\sum_{n=L+1}^{\infty} f_n \, H_n(Y_\alpha) = f(Y_\alpha) - f_0 - \sum_{n=1}^{L} f_n \, H_n(Y_\alpha)$$

So the estimator of $f[Y(x)]$ which has an infinity of terms:

$$f(Y(x))^* = f_0 + \sum_{n=1}^{L} f_n \sum_{\alpha} \lambda_{n\alpha} \, H_n(Y_\alpha) + \frac{1}{N} \sum_{\alpha} [\sum_{n=L+1}^{\infty} f_n \, H_n(Y_\alpha)]$$

becomes:

$$f_0 + \sum_{n=1}^{L} f_n \sum_{\alpha} \lambda_{n\alpha} \, H_n(Y_\alpha) + \frac{1}{N} \sum_{\alpha} [f(Y_\alpha) - f_0 - \sum_{n=1}^{L} f_n \, H_n(Y_\alpha)]$$

or more simply:

$$f(Y(x))^* = \frac{1}{N} \sum_{\alpha} f(Y_\alpha) + \sum_{n=1}^{L} f_n \sum_{\alpha} (\lambda_{n\alpha} - \frac{1}{N}) H_n(Y_\alpha).$$

There is now only a finite number of terms. And as:

$$\sum_{\alpha} (\lambda_{n\alpha} - \frac{1}{N}) = 1 - 1 = 0$$

we find again as expected:

$$E[f(Y(x))^*] = \frac{1}{N} \sum_{\alpha} E[f(Y_\alpha)] = E[f(Y(x))].$$

This was for a point estimation. It is also possible to introduce unbiasedness conditions in the DK estimation of regularized variables. But this will *not* be possible when dealing with a change of support (see Chapter 12).

Lognormal kriging

Adding unbiasedness conditions to remove local bias *cannot* be done in general when using the conditional expectation. The condition $\Sigma\, \lambda_\alpha = 1$ in the kriging system for the gaussian equivalents would not ensure the removal of local bias when estimating a non-linear function. A notable exception is the case where the function f is an exponential, which corresponds to estimating a lognormal variable. As a matter of fact, if $Z(x) = f[Y(x)]$ is a lognormal variable with log mean μ and log variance σ^2, then:

$$\log Z(x) = \log f[Y(x)] = \mu + \sigma\, Y(x)$$

and we can write:

$$Z(x) = f[Y(x)] = e^{\mu + \sigma Y(x)} = e^{\mu}\, e^{\sigma Y(x)}.$$

Its mean is (see Exercise 3, Chapter 5):

$$E[Z(x)] = E[f[Y(x)]] = e^{\mu}\, E[e^{\sigma Y(x)}] = e^{\mu + \sigma^2/2}$$

We can estimate $Z(x) = f[Y(x)]$ by integrating (7.1), which gives:

$$\exp\left(\mu + \sigma \sum \lambda^\alpha\, Y_\alpha + \sigma^2\, \sigma_K^2/2\right)$$

or, knowing $\log Z_a = \mu + \sigma Y_\alpha$:

$$\exp\Big((1 - \sum \lambda^\alpha)\mu + \sum \lambda^\alpha \log Z_\alpha + \sigma^2\, \sigma_K^2/2\Big)$$

$$= \exp\big([\log Z(x)]^{SK} + \sigma^2\, \sigma_K^2/2\big).$$

The term $\sigma^2\, \sigma_K^2$ (which must not be forgotten) is the kriging variance when estimating $\log Z(x)$. If γ denotes the variogram of the logarithms then:

$$\sigma^2\, \sigma_K^2 = \sigma^2 - \sum \lambda^\alpha \lambda^\beta (\sigma^2 \gamma_{\alpha\beta}) = [1 - (\sum \lambda^\alpha)^2]\sigma^2 + \sum \lambda^\alpha \lambda^\beta \gamma_{\alpha\beta}\,.$$

So the estimator becomes:

$$\exp[(1 - \sum \lambda^\alpha)\mu + \sum \lambda^\alpha \log Z_\alpha$$

$$+ (1 - (\sum \lambda^\alpha)^2)\frac{\sigma^2}{2} + \frac{1}{2}\sum \lambda^\alpha \lambda^\beta \gamma_{\alpha\beta}].$$

Now if we impose the condition $\Sigma\, \lambda_\alpha = 1$, we obtain:

$$Z(x)^* = \exp\Big(\sum \lambda^\alpha \log Z_\alpha + \frac{1}{2}\sum \lambda^\alpha \lambda^\beta \gamma_{\alpha\beta}\Big).$$

This estimator is unbiased whatever the local mean and variance, say μ' and σ'^2, of the logarithm, and hence whatever the local mean of the raw variable:

$$E[Z(x)^*] = \exp\Big(\mu' + \frac{1}{2}\sum \lambda^\alpha \lambda^\beta (\sigma'^2 - \gamma_{\alpha_\beta}) + \frac{1}{2}\sum \lambda^\alpha \lambda^\beta \gamma_{\alpha\beta}\Big)$$

$$= e^{\mu' + \sigma'^2/2} = E[e^{\log Z(x)}] = E[Z(x)].$$

Comments

Whether there is a condition or not on the λ_α, these estimators are obtained with the kriging (or another linear combination) of the logarithms. Be careful however! Just taking the expectation and forgetting the other terms is not enough and would introduce a dangerous bias. The term ‘lognormal kriging’ often covers such estimators.

10

Change of support in global estimation

Every geostatistician knows that the distribution of a variable, and in particular its variance, depends on the support on which the variable is defined. Often the data are values measured on an almost point support, whereas what we want to estimate is defined on a larger support. Hence the importance of models that can take the change of support into account.

We will see here a commonly used and very powerful model: the discrete gaussian model. We will consider the case of estimating the global distribution first (the present chapter), then (next chapter) the case of local estimation.

Introduction

In the previous chapters we have seen how to estimate a function of a point variable, given sample information at several other points. For example, we saw how to estimate an indicator:

$$I_{Z(x)<z} .$$

Its mean value over the region represents the cumulative distribution function of the variable $Z(x)$:

$$F(z) = P[Z(x) < z] = E[I_{Z(x)<z}].$$

To be precise, the point corresponds here to the support of the samples, and is usually very small, e.g. a drill core on which the metal grade has been measured. The grades measured on a small support like this can be much richer or poorer than those measured on larger supports v, e.g. mining blocks, which have more intermediate values (see Fig. 10.1). Of course, the average grade is the same whether we consider small supports or large ones. The difference in the distribution between large and small supports is called the support effect by geostatisticians.

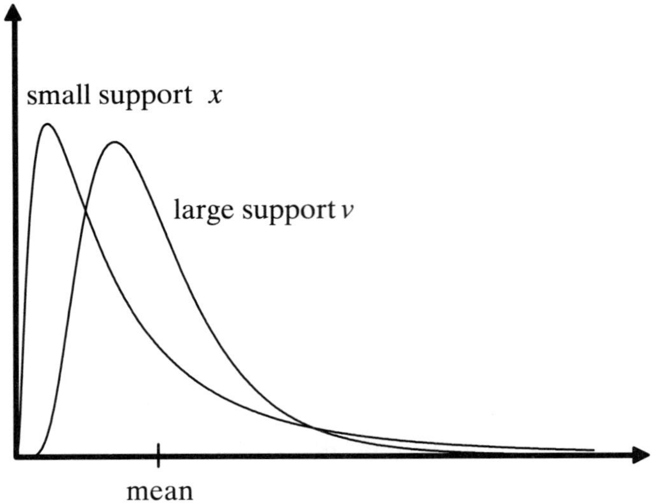

Fig. 10.1 Distribution of values

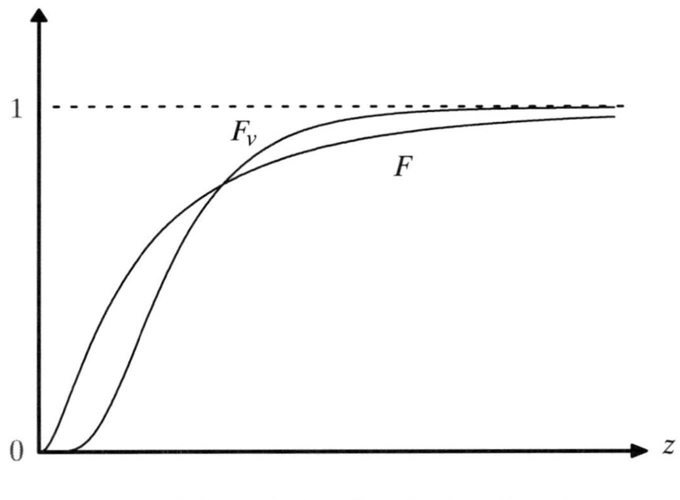

Fig. 10.2 Cumulative distribution functions

This difference also can be looked at on the cumulative distribution functions (Fig. 10.2). As the c.d.f. represent the means of the indicators $I(Z(x) < z)$ and $I(Z(v) < z)$

$$F(z) = E[I_{Z(x)<z}] = P[Z(x) < z]$$

$$F_v(z) = E[I_{Z(v)<z}] = P[Z(v) < z]$$

the difference shows that estimating $I(Z(x) < z)$ for a point support is not equivalent to estimating $I(Z(v) < z)$ for a larger block. A change of support model is needed to estimate the distribution of blocks given that of points.

In this work we assume that the histogram of our data (weighted, if necessary, to remove the effect of clustered data locations) is representative of the distribution of point values. But sometimes we cannot, particularly if there are too few samples or if their values are very variable. In these cases even the mean value over the region is not well known, and so it is better to compute the global estimation variance of this mean (linear geostatistics) or even to look for confidence intervals for the unknown mean value. In mining, this difficult problem has been studied by Sichel (1949, 1966) for independent and lognormally distributed values.

The aim of Chapters 10 to 12 is the change of support. This is vital for predicting recoverable reserves in mining, but it also plays a very important role in other fields such as environmental sciences where we have to estimate indicator functions or probabilities of exceeding thresholds.

Recoverable reserves

Let us now consider the miner's task of estimating recoverable reserves. Suppose that a block v has a grade $Z(v)$. Its recoverable reserves at cut-off z are:

- The ore tonnage, given apart from a multiplicative factor by the indicator $I(Z(v) < z)$.
- The quantity of metal, given by $Z(v)\ I(Z(v) \geq z)$.

For convenience we define the conventional profit function,[1] which is:

$$Z(v)\ I_{Z(v)\geq z} - z\ I_{Z(v)\geq z} = (Z(v) - z)\ I_{Z(v)\geq z}\,.$$

This is equal to $(Z(v) - z)$ if $Z(v)$ exceeds the cut-off z and is 0 otherwise.

[1] In effect the reader will show the following result. Imagine a mining exploitation where each of the blocks v (with known grade) can be selected independently from the others, where the value of a block is proportional to its metal content, and where the extraction and treatment costs, proportional to the tonnage, are the same for each block. With these conventions the profit is maximized by selecting each of the blocks if its metal pays these costs, i.e. if its grade exceeds a given cut-off z. For each of the blocks the profit is equal, apart from a constant factor, to our 'conventional profit'.

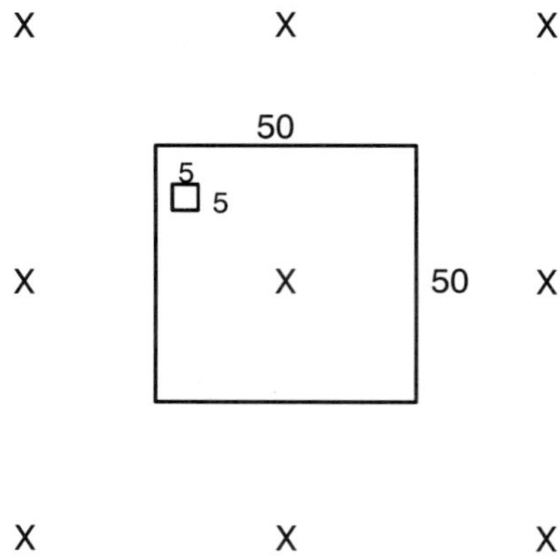

Fig. 10.3

Before mining commences we have data only from the exploration drill-holes. The grades $Z(v)$ of the blocks are not known. The recoverable reserves are precisely what we want to estimate in order to predict the results of selecting blocks above a certain cut-off. (We assume that selection is free; i.e. blocks will be selected individually as ore or waste independently of others.)

Suppose we have a 50×50 grid of sampling data. It would be pointless to estimate the reserves of each 5×5 selection block individually. When estimating the local reserves we should rather know the reserves contained in the $N = 100$ blocks v_i inside a 50×50 panel (Fig. 10.3). To be more precise we want to know:

- the ore tonnage of the panel (given by the number of blocks over the cut-off grade):

$$\sum I_{Z(v_i) \geq z}$$

  or equivalently, the proportion of these blocks in the panel:

$$\frac{1}{N} \sum I_{Z(v_i) \geq z}$$

- the metal tonnage:

$$\frac{1}{N} \sum Z(v_i)\, I_{Z(v_i) \geq z}.$$

For the global estimation of recoverable reserves our objective is to estimate these variables for all the blocks in the deposit that is:

- ore tonnage:

$$T(z) = E[I_{Z(v) \geq z}] = P[Z(v) \geq z]$$

- metal tonnage:

$$Q(z) = E[Z(v)\, I_{Z(v) \geq z}]$$

- conventional profit:

$$B(z) = E[(Z(v) - z)\, I_{Z(v) \geq z}] = Q(z) - z\, T(z)$$

- average recovered grade:

$$m(z) = \frac{Q(z)}{T(z)} = E[Z(v) | Z(v) \geq z].$$

Clearly if we know the distribution of block grades $Z(v)$ we can calculate all of these.

Remark: information effect and conditional bias

The recoverable reserves as they have been defined above assume that the block grades $Z(v)$ are known (exactly) when the selection is being made. So they take account of the support effect.

It is true that the ultimate information available when the selection is made will be much more detailed than at the exploration stage, but the block grade $Z(v)$ will still not be known exactly. So the selection will be based on estimated values, and consequently it will not be as good as if the grades were known. The estimates of recoverable reserves should also take account of this information effect as it is called.

Suppose that the ultimate estimator $Z(v)^{\mathrm{ult}}$ is 'conditionally unbiased' (which is practically the case with kriging if its neighbourhood is large enough); that is,

$$E[Z(v) | Z(v)^{\mathrm{ult}}] = Z(v)^{\mathrm{ult}}.$$

The average recovered grade:

$$E[Z(v) | Z(v)^{\mathrm{ult}} \geq z]$$

will equal the predicted value:

$$E[Z(v)^{\mathrm{ult}} | Z(v)^{\mathrm{ult}} \geq z].$$

We then need to know the distribution of the final estimated grades $Z(v)^{ult}$ (instead of the real block grades $Z(v)$) in order to estimate the reserves.

Selectivity curves

In mining the distribution of grades is often represented by the corresponding 'grade-tonnage' curves. For example mine surveyors plot:

- the ore tonnage T as a function of the cut-off grade z
- the metal tonnage Q as a function of the cut-off grade z
- the average grade m as a function of the cut-off grade z
- the conventional profit B as a function of the cut-off grade z
- the average grade m as a function of the ore tonnage T
- the metal tonnage Q as a function of the ore tonnage T

The functions $T(z)$, $Q(z)$, $B(z)$, and $m(T)$ are always decreasing functions whereas $m(z)$ and $Q(T)$ are increasing functions. Two of these functions in particular (the metal against ore tonnage and the conventional profit against cut-off) are valuable when one wants to compare the selectivity for different-sized supports (Matheron 1981, 1984).

The graph of metal against ore tonnage is always convex from below (Fig. 10.4). If we draw a line from any point (T, Q) on this curve to the origin, its slope is $Q/T = m$. The slope of the tangent at the point (T, Q) is equal to the cut-off z, and the tangent cuts the Q axis at the value B.

Now if we plot the metal against ore tonnage for two different supports, the one for the small support always lies above the one for the larger support for the same ore tonnage. Fig. 10.5 shows this.

The other profitable curve is the one for conventional profit against cut-off (Fig. 10.6). This is always convex. The tangent at the point (B, z) cuts the two axes at the points (O, Q) and $(m = Q/T, O)$, and its slope is $-T$.

If we plot these curves for several supports we see that for a given cut-off grade the conventional profit decreases as the size of support increases (Fig. 10.7).

Plotting the sets of curves of $Q(T)$ and of $B(z)$ is very valuable for assessing and comparing the selectivity at different support sizes. Clearly, the change of support models used to estimate the distribution of block grades from that of the samples must respect the ordered relations among the curves.

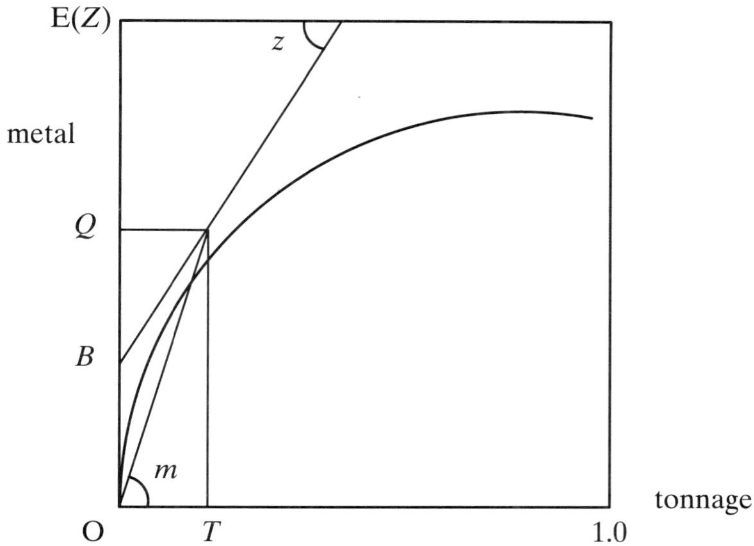

Fig. 10.4

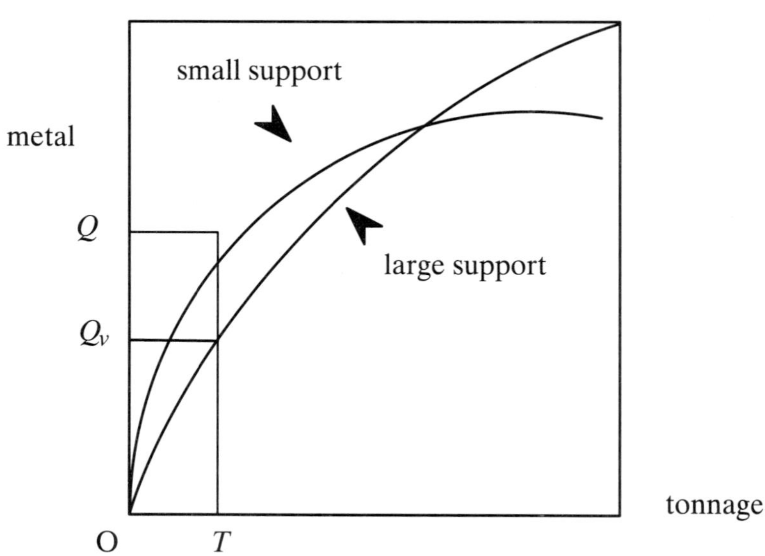

Fig. 10.5

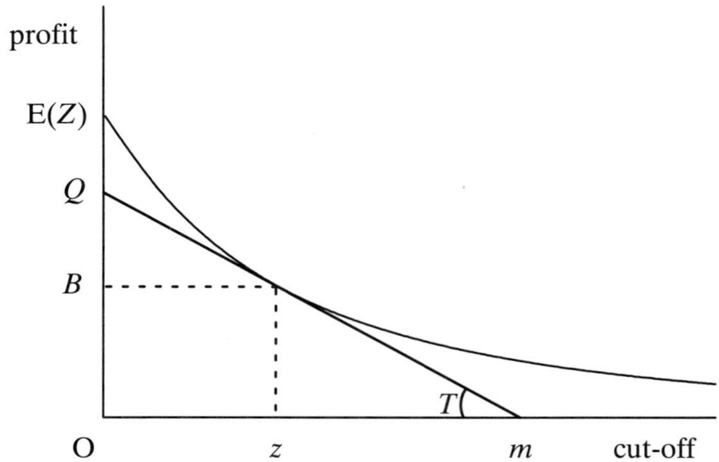

Fig. 10.6

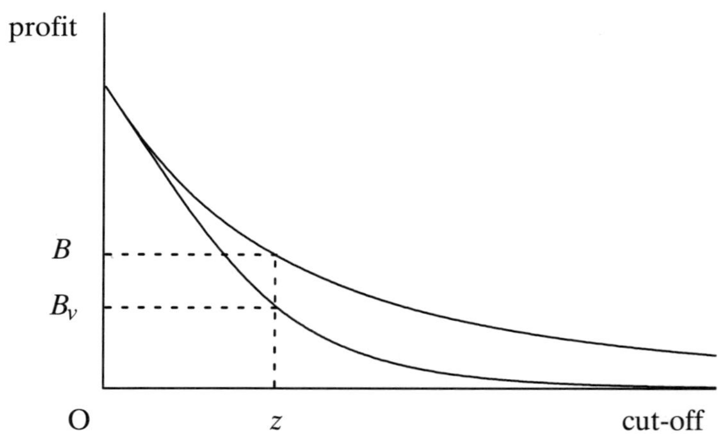

Fig. 10.7

Cartier's relation

We now want to work out how to obtain the distribution of the blocks $Z(v)$ from that of the grades $Z(x)$ (which we know). We need a model to do this, and it has to satisfy certain conditions.

First, the mean values must be the same:

$$E[Z(x)] = E[Z(v)].$$

Second, from linear geostatistics we can calculate the variance of $Z(v)$ from the variogram model $\gamma(h)$ of $Z(x)$:

$$\text{var } Z(v) = \text{var } Z(x) - \bar{\gamma}(v, v).$$

The term $\bar{\gamma}(v, v)$ represents the within block variability, i.e. the dispersion variance of points inside a block v. It is the mean value of $\gamma(x - y)$ where x and y describe the block independently:

$$\bar{\gamma}(v, v) = \frac{1}{v^2} \int\int \gamma(x - y)\, dx\, dy.$$

If there is a large nugget effect or a large short-scale structure, $\bar{\gamma}(v, v)$ will be large, and so the support effect is marked.

Third, there is Cartier's relation:

$$E[Z(\underline{x}) \mid Z(v)] = Z(v)$$

where $\underline{x}$ denotes a point located at random inside the block v. This condition means that on average the expected value of the grade of a point chosen at random given the block grade is just equal to the block grade. It is important because we can show (cf. Matheron 1981) that it guarantees the order relationship seen in the metal against ore tonnage curve and in the conventional profit against cut-off grade curve.

The discrete gaussian model (1)

Preliminary remark. In the discrete gaussian model, the field is considered as the union of blocks, and the samples are randomly located within their block. In the narrow sense this model can be used for a change of support for the global distribution (this chapter). In the wider sense (see next chapter) it is a more complete model that can be used for the local estimation of distributions.

In the gaussian case, the point variable $Z(x)$ can be expressed as a function of a standard normal variable $Y(x)$:

$$Z(x) = \Phi(Y(x)).$$

The function Φ is the gaussian anamorphosis function that we described in Chapters 5 and 6. It is known experimentally because we know the distribution of $Z(x)$.

The variable $Z(v)$ can also be expressed as a function Φ_v of another standard normal variable Y_v:

$$Z(v) = \Phi_v(Y_v).$$

We now will determine Φ_v, which is equivalent to determining the distribution of $Z(v)$. We can rewrite Cartier's relation:

$$E[\Phi(Y(\underline{x}))\,|\,\Phi_v(Y_v)] = \Phi_v(Y_v)$$

or

$$\Phi_v(Y_v) = E[\Phi(Y(\underline{x}))\,|\,Y_v].$$

To evaluate this we make one more hypothesis. We assume that the pair $(Y_v, Y(\underline{x}))$ has a bivariate normal distribution with coefficient of correlation r. Since Y_v and $Y(\underline{x})$ are standard normal variables the conditional distribution of $Y(\underline{x})$ given $Y_v = y_v$ is normal with mean ry_v and variance $1 - r^2$ (see 5.9):

$$Y(\underline{x}) = ry_v + \sqrt{1 - r^2}\,U.$$

This gives us the anamorphosis function Φ_v for blocks:

$$\Phi_v(Y_v) = \int \Phi(r\,Y_v + \sqrt{1 - r^2}\;u)\;g(u)\;du \qquad (10.1)$$

where $g(u)$ is the density function of the standard normal distribution.[1]

In practice it is convenient to use Hermite polynomials. As Y_v and $Y(\underline{x})$ are standard normal variables, a relation similar to (5.3) gives:

$$E[H_n(Y(\underline{x}))\,|\,Y_v] = r^n\;H_n(Y_v).$$

So:

$$\begin{aligned}\Phi_v(Y_v) &= E[\Phi(Y(\underline{x}))\,|\,Y_v] \\ &= E[\sum_0 \phi_n\;H_n(Y(\underline{x}))\,|\,Y_v] \\ &= \sum_0 \phi_n\;E[H_n(Y(\underline{x}))\,|\,Y_v]\end{aligned}$$

[1] This change of support formula conserves lognormality: if $Z(x)$ is lognormal so is $Z(v)$. Proof: We know (5.9) that a lognormal variable $Z(x)$ with mean $M = E(Z)$ can be written as:

$$Z(x) = M\,\exp(\sigma\;Y(x) - \sigma^2/2).$$

Using formula (10.1), we obtain:

$$Z(v) = M\,\exp(r\;\sigma\;Y_v - r^2\;\sigma^2/2).$$

This is also a lognormal variable with same mean M but its logarithmic variance $r^2\,\sigma^2$ is smaller than the logarithmic variance σ^2 of $Z(x)$.

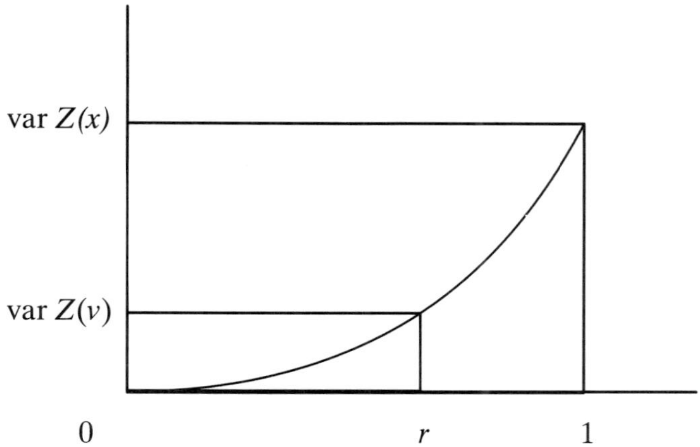

Fig. 10.8

that is:

$$Z(v) = \Phi_v(Y_v) = \sum_0 \phi_n \, r^n \, H_n(Y_v). \qquad (10.2)$$

So we obtain the anamorphosis coefficients for blocks by multiplying the corresponding point coefficients ϕ_n by r^n. We note that the lowest-order term ϕ_0 is the same since:

$$\phi_0 = E[Z(x)] = E[Z(v)].$$

As there is no correlation between the different polynomials, we obtain (as in Chapter 5):

$$\text{var } Z(v) = Var \, \Phi_v(Y_v) = \sum_1 (\phi_n)^2 \, r^{2n}.$$

Since var $Z(v) = Var\ Z(x) - \bar{\gamma}\,(v, v)$, we can compute the value of the change of support coefficient r which corresponds to the known variance of blocks (see Fig. 10.8). Note $0 \leq r \leq 1$.

Now that r is known, so is the block anamorphosis function Φ_v and hence the block distribution. Since Y_v is a standard normal variable (density g and cumulative distribution G), we can easily estimate any function of $Z(v)$, or equivalently any function of Y_v. As:

$$Z(v) = \Phi_v(Y_v) \geq z \quad \Leftrightarrow \quad Y_v \geq y$$

with $z = \Phi_v(y)$, we see that for example:

$$F_v(z) = E[I_{Z(v)<z}] = P(Z(v) < z)$$
$$= E[I_{Y_v<y}] = P(Y_v < y) = G(y).$$

So we can estimate the global recoverable reserves of a deposit when selecting blocks at cut-off z. The ore tonnage and quantity of metal are given by:

$$T_v(z) = E[I_{Z(v)\geq z}] = 1 - G(y)$$
$$Q_v(z) = E[Z(v)\ I_{Z(v)\geq z}]$$
$$= E[\Phi_v(Y_v)\ I_{Y_v\geq y}]$$
$$= \int_y \Phi_v(y)\ g(y)\ dy.$$

We can calculate $Q_v(z)$ using Hermite polynomials:

$$Q_v(z) = \int_y (\sum_0 \phi_n\ r^n\ H_n(y))\ g(y)\ dy$$
$$= \sum_0 \phi_n\ r^n \int_y H_n(y)\ g(y)\ dy.$$

As in Chapter 6, we can show that this is equal to:

$$\phi_0\ [1 - G(y)] - \sum_1 \phi_n\ r^n \frac{1}{\sqrt{n}}\ H_{n-1}(y)\ g(y).$$

Exercises

Here are some simple exercises. In these $Z(x)$ is a stationary variable with mean m and with covariance $C(h)$. So its variance is $C(0)$, and its variogram is:

$$\gamma(h) = C(0) - C(h).$$

Exercise 1

Let $Z(v)$ be the average for a block v

$$Z(v) = \frac{1}{v} \int_v Z(x)\ dx.$$

Let x_0 be a fixed point inside v. Its value is $Z(x_0)$.

(a) Suppose that we know $Z(x_0)$ and want to estimate $Z(v)$. We will consider successively the cases where the mean m is unknown (ordinary kriging) and where the mean is known (simple kriging). Show that the ordinary kriging reduces to:

$$Z(v)^{OK} = Z(x_0)$$

and show that the simple kriging can be written:

$$Z(v)^{SK} = \lambda\ Z(x_0) + (1 - \lambda)\ m$$

with the weight $\lambda = C(x_0, v)/C(0)$ given to $Z(x_0)$ and $1 - \lambda$ given to the mean.

(b) Now suppose we know $Z(v)$ and want to estimate $Z(x_0)$ as before using ordinary kriging and simple kriging. Show that ordinary kriging is:

$$Z(x_0)^{OK} = Z(v)$$

and that simple kriging can be written:

$$Z(x_0)^{SK} = \lambda\ Z(v) + (1 - \lambda)\ m$$

with the weight $\lambda = C(x_0, v)/C(v, v)$ given to $Z(v)$ and $1 - \lambda$ given to the mean.

(c) This time, repeat question (a) for a point $\underline{x}$ located at random inside v. Show that the weight of $Z(\underline{x})$ in SK now equals $\lambda = C(v, v)/C(0)$.

(d) Repeat question (b) for a point $\underline{x}$ located at random inside v. Show that the weight given to $Z(v)$ in SK is $\lambda = C(v, v)/C(v, v) = 1$ and that given to the mean is zero. Deduce that when a sample $Z(\underline{x})$ is taken at random inside the block v, the kriged estimator of its value, given the block grade, is equal to the block grade. Does this result agree with Cartier's relation:

$$E[Z(\underline{x})\,|\,Z(v)] = Z(v)\ ?$$

Exercise 2

Now we use the gaussian change of support model. As before v denotes the block and $\underline{x}$ denotes a point taken at random inside v.

$$Z(v) = \Phi_v\,[Y_v] = \sum_0 \phi_n\ r^n\ H_n(Y_v)$$

$$Z(\underline{x}) = \Phi[Y(\underline{x})] = \sum_{0} \phi_n \, H_n(Y(\underline{x})).$$

The pair $(Y_v, Y(\underline{x}))$ is bivariate normal with a correlation coefficient equal to the change of support coefficient r. By construction this model satisfies Cartier's relation.

(a) Suppose that we know the value of $Z(\underline{x})$. Express the expected value of $Z(v)$ given $Z(\underline{x})$ in terms of Hermite polynomials:

$$E[Z(v)\,|\,Z(\underline{x})] = \sum_{0} \phi_n \, r^{2n} \, H_n(Y(\underline{x})).$$

Hint: as Y_v and $Y(\underline{x})$ are standard bivariate normal with correlation coefficient r, we know that (as in (5.3)):

$$E[H_n(Y_v)\,|\,Y(\underline{x})] = r^n \, H_n[Y(\underline{x})].$$

(b) As in (a) we know $Z(\underline{x})$. We want to estimate $I(Z(v) < z)$ given $Z(\underline{x})$ using its expected value:

$$E[I_{Z(v)<z}\,|\,Z(\underline{x})] = P[Z(v) < z\,|\,Z(\underline{x})].$$

Show that this can be expressed simply in terms of the cumulative normal distribution G:

$$G\left(\frac{y - r\,Y(\underline{x})}{\sqrt{1 - r^2}}\right)$$

where y satisfies $z = \Phi_v(y)$.

Hint: since Y_v and $Y(\underline{x})$ are standard bivariate normal with correlation coefficient r, the distribution of Y_v knowing $Y(\underline{x}) = y(x)$ is normal with mean $ry(x)$ and variance $1 - r^2$. So we can write it as:

$$Y_v = ry(x) + \sqrt{1 - r^2}\, U$$

where U is a standard normal variable.

(c) Now suppose we know $Z(v)$ and want to estimate $I(Z(\underline{x}) < z)$. Show that the estimator can be written:

$$G\left(\frac{y - r\,Y_v}{\sqrt{1 - r^2}}\right)$$

where y now satisfies $z = \Phi(y)$.

11

Change of support in local estimation

In the previous chapter we have seen how the global estimation of distributions can be performed using the gaussian discrete model. We shall now see how this model makes the local estimation possible, whether by DK or by conditional expectation.

A point-block model

Suppose that we have a certain number of point samples $Z(x_\alpha)$ and that we want to estimate a function of the average grade of a particular block $Z(v)$. For example the function could be an indicator:

$$I_{Z(v)<z} .$$

In the first half of this book, we were looking for models that linked the values observed at different points (i.e. a point-point model). We now need a model linking the point values with those of blocks. This problem is theoretically very complex, and so we need to make some simplifying assumptions. The procedure used is described below for the gaussian case but it can be extended to others (e.g. gamma).

(a) The region is divided into blocks v_i whose size is equal to the support v. For example, in mining these blocks are the mining selection units.

  Remark. Here, the block size and the position of the blocks are fixed (or nearly fixed). A high-grade area of limited size that would be found during mining may therefore be included in a single block, or split into two or more blocks. In reality it is also possible for the geometry of mining blocks to differ from block to block, and to depend on the location of the ore found during mining. This case is called 'adaptative' geometry. There are no models for predicting directly recoverable reserves in this case, so a simulation must be used.

(b) We assume that the position of a sample inside a block is random, and consequently 'forget' its exact position.

(c) We assume that given the grade $Z(v_i)$ of the block v_i containing a particular sample x_i, the sample grade $Z(x_i)$ is independent of all the other grades, blocks or points.

Now we are going to see what can be said about the different covariances. To start with our point support variable $Z(x)$ has a variogram $\gamma(h)$ (or a covariance $C(h)$) that is assumed to be known. From linear geostatistics, we know that the covariance between two blocks v_i and v_j , denoted by $C(v_i, v_j)$, is the mean value of $C(h)$ when one end of the vector h sweeps over v_i while the other end independently sweeps over v_j. So we can evaluate the covariance between any two blocks.

Similarly we can calculate the covariance between a point x_i and a block v_j. So now we let x_i be a point chosen at random in v_i. Clearly the covariance between x_i and v_j written as $C(x_i, v_j)$ is just the covariance between v_i and v_j, that is $C(v_i, v_j)$.

In the same way, if x_i and x_j are points chosen at random in their respective blocks then the covariance $C(x_i, x_j)$ equals $C(v_i, v_j)$ too. Note that if $v_i = v_j = v$ this covariance $C(v, v)$ which is equal to the variance of the blocks also represents the covariance between two samples chosen at random in a block. In contrast to this, the covariance between a sample and itself (even if the sample is selected at random) is always equal to $C(0)$, that is, the variance of the samples.

Except for this last remark, we see that the point-point and point-block covariances built up in this model by randomizing the samples within their blocks are just equal to the block-block covariance which is considered as known.

The discrete gaussian model (2)

We associate the corresponding gaussian value Y_{v_i} with each block grade $Z(v_i)$, in the same way that we associated a gaussian equivalent $Y(x_i)$ with each sample grade $Z(x_i)$. In the previous chapter we saw that if we know the coefficients of the Hermite polynomials ϕ_n for the point anamorphosis, we can deduce those for the block anamorphosis Φ_v: they are just $\phi_n r^n$. The change of support coefficient r was calculated so that the block variance corresponds to the correct value.

In order to link the different blocks and samples we are now going to assume that any set of point gaussian equivalents $Y(x_i)$, $Y(x_j)$, . . . and of block gaussian equivalents Y_{v_i}, Y_{v_j}, . . . is multivariate normal. As these gaussian equivalents have zero mean and unit variance, the model is fully determined once we know the covariance between them. We now work this out.

Let $\rho_{v_i\,v_j}$ be the covariance (or the correlation) between Y_{v_i} and Y_{v_j}. As the corresponding raw variables $Z(v_i)$ and $Z(v_j)$ have $\phi_n\,r^n$ as the coefficients of their anamorphosis, we can check, following (6.2), that:

$$C(v_i, v_j) \;=\; \sum_{1} \phi_n^2\, r^{2n}\, (\rho_{v_i\,v_j})^n. \qquad (11.1)$$

In practice, this equation is inverted to give us the covariance $\rho_{v_i\,v_j}$ once we know $C(v_i, v_j)$, ϕ_n, and r.

Now let $\rho_{x_i\,v_j}$ be the covariance (or the correlation) between the gaussian equivalent $Y(x_i)$ and Y_{v_j}. This time, the coefficients of the anamorphosis for $Z(x_i)$ are ϕ_n, while those for $Z(v_j)$ are $\phi_n\,r^n$. As before we can show that:

$$C(x_i, v_j) \;=\; \sum_{1} \phi_n^2\, r^n\, (\rho_{x_i\,v_j})^n.$$

Now since $C(x_i, v_j)$ is equal to $C(v_i, v_j)$ we obtain:

$$C(v_i, v_j) \;=\; \sum_{1} \phi_n^2\, r^n\, (\rho_{x_i\,v_j})^n.$$

By comparing this equation with (11.1), we can deduce the relation between the point-block gaussian covariance and the block-block one:

$$\rho_{x_i\,v_j} \;=\; r\, \rho_{v_i\,v_j} \qquad (11.2)$$

That is, one is obtained from the other by multiplying by r.

Lastly, we can also show (6.2) that the covariance $\rho_{x_i\,x_j}$ between $Y(x_i)$ and $Y(x_j)$ satisfies:

$$C(x_i,\, x_j) \;=\; \sum_{1} \phi_n^2\, (\rho_{x_i\,x_j})^n.$$

So:

$$C(v_i, v_j) \;=\; \sum_{1} \phi_n^2\, (\rho_{x_i\,x_j})^n.$$

Similarly, by comparison with (11.1), $\rho_{x_i\,x_j}$ is obtained from $\rho_{v_i\,v_j}$ by multiplying by r^2:

$$\rho_{x_i\,x_j} \;=\; r^2\, \rho_{v_i\,v_j}. \qquad (11.3)$$

Thus the discrete gaussian model is fully determined once we know the change of support coefficient r and the covariance between the block gaussian equivalents $\rho_{v_i\,v_j}$.

In practice we proceed as follows. We regularize the covariance $C(h)$ of the raw variable $Z(x)$ to get $C(v_i, v_j)$ for the different possible distances between v_i and v_j. We can then obtain the values of $\rho_{v_i\, v_j}$ for these distances by inverting the formula given above. These values are then fitted into a covariance model. Once we know this covariance, the change of support coefficient, and the Hermite polynomial coefficients, the discrete gaussian model is fully specified.

So it is a simple, neat, and yet mathematically consistent model. Moreover it turns out to be very useful in practice in non-linear geostatistics. In the following pages we shall see how it can be used for local estimation using bigaussian DK or multinormal conditional expectation. Moreover, it can also be used for non-conditional simulations and, thanks to the valuable properties of the multinormal distribution, conditional simulations.

Exercise

When introducing the point-block model we assumed (point (c)) that, knowing its block value, the value of a sample is independent of other block and point values. We now verify, on a particular example, that this hypothesis is satisfied in the gaussian discrete model through the relations between the covariances of the gaussian equivalents.

Let us take two blocks v_i and v_j and a sample x_i inside v_i. Let $Z(v_i)$, $Z(v_j)$, and $Z(x_i)$, be the corresponding grades and Y_{v_i}, Y_{v_j}, and $Y(x_i)$ their gaussian equivalents. We will show that, knowing the grade of its block, the grade of the sample x_i is independent of the other block, or equivalently, that, knowing Y_{v_i}, $Y(x_i)$ is independent of Y_{v_j}.

(a) Show that $Y(x_i)$ can be written as:

$$Y(x_i) = r\, Y_{v_i} + \sqrt{1 - r^2}\; U_1$$

with U_1 standard normal variate independent of Y_{v_i}, and that in particular, knowing $Y_{v_i} = y_{v_i}$, $Y(x_i)$ is equal to:

$$Y(x_i) = r\, y_{v_i} + \sqrt{1 - r^2}\; U_1.$$

Hint: look at Exercise 2, Chapter 5 and use the fact that Y_{v_i} and $Y(x_i)$ are a standard normal pair with correlation r.

(b) In the same way show that Y_{v_j} can be written as:

$$Y_{v_j} = \rho_{v_i\, v_j}\, Y_{v_i} + \sqrt{1 - \rho_{v_i\, v_j}^{\;2}}\; U_2$$

with U_2 standard normal variate independent of Y_{v_i}, and that in particular, knowing $Y_{v_i} = y_{v_i}$, Y_{v_j} is equal to:

$$Y_{v_j} = \rho_{v_i v_j} \, y_{v_i} + \sqrt{1 - \rho_{v_i v_j}{}^2} \; U_2.$$

(c) From the previous relations deduce that the covariance between $Y(x_i)$ and Y_{v_j} equals:

$$\text{cov } (Y(x_i), \; Y_{v_j}) = E(Y(x_i) \; Y_{v_j})$$

$$= r \, \rho_{v_i v_j} + \sqrt{1 - r^2} \sqrt{1 - \rho_{v_i v_j}{}^2} \, E(U_1 U_2)$$

but that, conditional on $Y_{v_i} = y_{v_i}$, the covariance between $Y(x_i)$ and Y_{v_j} is reduced to:

$$\sqrt{1 - r^2} \sqrt{1 - \rho_{v_i v_j}{}^2} \; E(U_1 U_2).$$

(d) From (11.2) the covariance between $Y(x_i)$ and Y_{v_j} in the discrete gaussian model equals $r \, \rho_{v_i v_j}$. Deduce:

$$E(U_1 U_2) = 0.$$

Conditional on $Y_{v_i} = y_{v_i}$, the covariance between $Y(x_i)$ and Y_{v_j} is then zero, so that these normal variables are independent as expected.

Disjunctive kriging

In the discrete gaussian model, the multivariate distributions of the gaussian variables Y_{v_i}, Y_{v_j}, . . . $Y(x_i)$, $Y(x_j)$, . . . are multivariate normal. In particular, the bivariate distributions are bivariate normal. This allows us to use disjunctive kriging. The problem is to use the appropriate covariance models.

Suppose that we want to estimate the indicator functions $I(Z(v) < z)$ for a block v given the neighbouring data $Z(x_\alpha)$. This comes back to estimating $I(Y_v < y)$ given $Y(x_\alpha)$, where $Z(v) = \Phi_v(Y_v)$ and $z = \Phi_v(y)$. Now $I(Y_v < y)$ is a function of Y_v and it can be expanded in terms of Hermite polynomials $H_n(Y_v)$. This was given in (5.2):

$$I_{Y_v < y} = G(y) + \sum_1 \frac{1}{\sqrt{n}} H_{n-1}(y) \, g(y) \, H_n(Y_v).$$

In order to estimate this by disjunctive kriging, we have to krige each $H_n(Y_v)$ given $H_n(Y_\alpha)$ using simple kriging. That is:

$$[I_{Y_v < y}]^{DK} = G(y) + \sum_1 \frac{1}{\sqrt{n}} H_{n-1}(y) \, g(y) \, [H_n(Y_v)]^K.$$

We need to krige each of the $H_n(Y_v)$ only once to obtain the DK estimator of any other function of Y_v or of $Z(v)$ that we want to estimate.

The kriging of $H_n(Y_v)$ with a known mean (equal to zero) can be written as:

$$[H_n(Y_v)]^K = \sum_{\alpha} \lambda_{n\alpha} H_n(Y_\alpha)$$

but it is more difficult to write the corresponding system. To start with, we write it in general form:

$$\sum_{\beta} \lambda_{n\beta} \operatorname{cov} [H_n(Y_\alpha), H_n(Y_\beta)] = \operatorname{cov} [H_n(Y_\alpha), H_n(Y_v)] \qquad \equiv \alpha$$

Then we note that as the distributions are bivariate normal (5.4),

$$\operatorname{cov} [H_n(Y_\alpha), H_n(Y_\beta)] = [\operatorname{cov} (Y_\alpha, Y_\beta)]^n$$

and

$$\operatorname{cov} [H_n(Y_\alpha), H_n(Y_v)] = [\operatorname{cov} (Y_\alpha, Y_v)]^n.$$

So the kriging system is:

$$\sum_{\beta} \lambda_{n\beta} [\operatorname{cov} (Y_\alpha, Y_\beta)]^n = [\operatorname{cov} (Y_\alpha, Y_v)]^n \qquad \text{for all } \alpha.$$

As we are using the discrete gaussian model we know that:

$$\begin{aligned}\operatorname{cov} (Y_\alpha, Y_v) &= r \operatorname{cov} (Y_{v_\alpha}, Y_v)\\ &= r\, \rho_{v_\alpha v}\end{aligned}$$

and:

$$\begin{aligned}\operatorname{cov} (Y_\alpha, Y_\beta) &= r^2 \operatorname{cov} (Y_{v_\alpha}, Y_{v_\beta})\\ &= r^2\, \rho_{v_\alpha v_\beta}\end{aligned}$$

except for $\alpha = \beta$:

$$\operatorname{cov} (Y_\alpha, Y_\alpha) = \operatorname{var} Y_\alpha = 1.$$

So the system turns out to be:

$$\lambda_{n\alpha} + \sum_{\alpha \neq \beta} \lambda_{n\beta}\, r^{2n} (\rho_{v_\alpha v_\beta})^n = r^n (\rho_{v_\alpha v})^n \qquad \text{for all } \alpha.$$

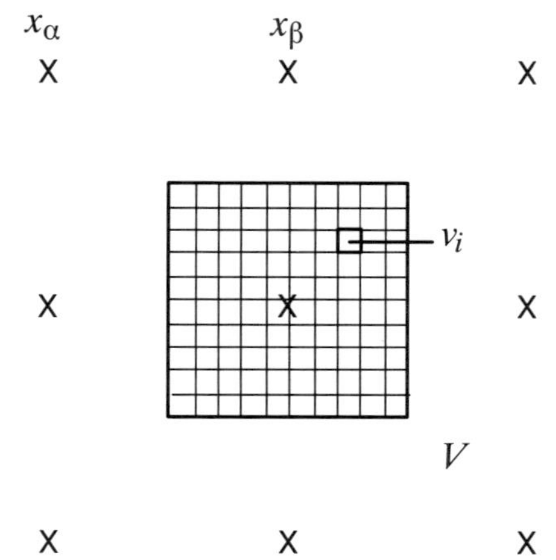

Fig. 11.1

As n increases, the structure of H_n tends to pure nugget; the kriging weights tend to 0 and so $[H_n(Y_v)]^K$ tends to its mean (zero) rapidly. Consequently only a few kriging systems have to be solved in practice.

In mining for example, we do not try to estimate the recoverable reserves for each small block, e.g. $I(Z(v) \geq z)$. We are satisfied with estimating the reserves corresponding to the N small blocks v_i inside a panel V which is the same size as the sample grid (Fig. 11.1). The ore tonnage recovered at a cut-off grade of z is then equal to the number of blocks above the cut-off:

$$\sum_i I_{Z(v_i) \geq z}.$$

It is more convenient to express this as a proportion of the total:

$$\frac{1}{N} \sum_i I_{Z(v_i) \geq z}.$$

The quantity of metal contained in this is:

$$\frac{1}{N} \sum_i Z(v_i)\, I_{Z(v_i) \geq z}.$$

It is not necessary to estimate $H_n(Y_{v_i})$ for each small block in order to get the DK estimator of this. As we saw earlier in Chapter 8:

$$\frac{1}{N} \sum_i I_{Z(v_i) \geq z} = 1 - G(y) - \sum_1 \frac{H_{n-1}(y)\, g(y)}{\sqrt{n}} \left[\frac{1}{N} \sum_i H_n(Y_{v_i}) \right].$$

So we need krige only $(1/N)\ \Sigma\ H_n(Y_{v_i})$ directly. The kriging system is then:

$$\lambda_{n\alpha} + \sum_{\beta \neq \alpha} \lambda_{n\beta}\, r^{2n}\, (\rho_{v_\alpha v_\beta})^n = \frac{1}{N} \sum_i r^n\, (\rho_{v_\alpha v_i})^n \qquad \text{for all } \alpha.$$

Note: the right-hand side is the mean of the nth power of $\rho_{v_\alpha v_i}$ for all the small blocks v_i, and *not* the nth power of the mean of the $\rho_{v_\alpha v_i}$!

Conditional expectation

As all the multivariate distributions considered in the discrete gaussian model are normal, this model can also be used for the conditional expectation estimator. Given the $Y(x_\alpha) = y_\alpha$, the variable Y_v is normal. Its mean is the simple kriging of y_v:

$$(y_v)^K = \sum \lambda_\alpha\, y_\alpha$$

and its variance is the kriging variance of y_v:

$$\sigma_K^2.$$

A similar reasoning to the one in Chapter 7, gives us:

$$\begin{aligned}[I_{Z(v)<z}]^{CE} = [I_{Y_v<y}]^{CE} &= E[I_{Y_v<y} \,|\, Y(x_\alpha) = y_\alpha \ldots] \\ &= P(Y_v < y \,|\, Y(x_\alpha) = y_\alpha \ldots) \\ &= G\left[\frac{y - \sum \lambda_\alpha\, y_\alpha}{\sigma_K}\right].\end{aligned}$$

This estimator has one important advantage compared to the disjunctive kriging one: it is an increasing function of z and varies from 0 to 1. To obtain it we only need to solve the kriging system for Y_v given the Y_α:

$$\sum_\beta \lambda^\alpha \operatorname{cov}(Y_\alpha, Y_\beta) = \operatorname{cov}(Y_\alpha, Y_v) \qquad \text{for all } \alpha.$$

Using the discrete gaussian model this becomes:

$$\lambda_\alpha + \sum_{\beta \neq \alpha} \lambda_\beta\, r^2\, \rho_{v_\alpha v_\beta} = r\, \rho_{v_\alpha v} \qquad \text{for all } \alpha.$$

In DK we were able to estimate the recoverable reserves inside a panel. If we now want to do that, in principle, we need to do the estimation block by block, that is

$$\left[\frac{1}{N}\sum_i I_{Z(v_i)\geq z}\right]^{CE} = \frac{1}{N}\sum [I_{Z(v_i)\geq z}]^{CE} = \frac{1}{N}\sum [I_{Y_{v_i}\geq y}]^{CE}.$$

This means solving the kriging system for each of the N small blocks. However, as the data used, the Y_α, are the same for all of the small blocks v_i inside the panel, the matrix on the left-hand side is always the same. So we need to invert the matrix only once.

Uniform conditioning using a normal variable

In the previous estimator, each of the N blocks Y_{v_i} inside a panel was conditioned by its kriged value with weights λ_α which vary from block to block. We can simplify this procedure by conditioning each of these blocks with a single linear combination $\Sigma\ \lambda_a\ Y_a$. For example, this combination could be the kriged value of $(1/N)\ \Sigma\ Y_{v_i}$. Using the discrete gaussian model we can calculate the required covariances, and hence the expression for the estimators. This is one type of 'uniform conditioning' (i.e. the conditioning is done by a *single* variable, which is here $\Sigma\ \lambda_a\ Y_a$).

There are other ways of uniform conditioning. We could, for example, estimate the recoverable reserves of a block $\underline{v}$ random inside V. In that case the conditioning is done by the kriged value of $Y_{\underline{v}}$.

As computers have become more powerful these types of uniform conditioning are no longer as valuable a simplification of the conditional expectation as they were at one time. The choice nowadays is between conditional expectation and disjunctive kriging. In practice, these two estimators are approximately equivalent. The conditional expectation, which is based on an assumption of multivariate normality as opposed to bivariate normality for disjunctive kriging, has the advantage of giving results that are completely consistent. However, if we do not make the assumption of normality, there is no convenient way of handling the multivariate distributions needed to evaluate the conditional expectation. The strong point of disjunctive kriging is precisely that it can be used for bivariate distributions other than the normal (e.g. gamma, negative binomial, with orthogonal indicator residuals, corresponding to the mosaic model). For example, we can construct a discrete gamma model in just the same way as we did for the discrete gaussian model.

12

Change of support and stationarity

When a change of support has to be taken into account, the direct estimation using gaussian DK or conditional expectation requires a strict hypothesis of stationarity. When this is too strong, the special methods presented in this chapter can be used: uniform conditioning by an imposed value, changing of support at sample points.

Disjunctive kriging and conditional expectation

The problems with stationarity mentioned in Chapter 9 in connection with point estimation also arise for estimates that require a change of support. Except for the lognormal case a non-bias condition could not solve the problems for the conditional expectation. It cannot solve it either when a change of support is involved.

Let us now consider DK. For example, in the gaussian case, when we want to estimate a point function:

$$f[Y(x)] = f_0 + \sum_{1} H_n[Y(x)]$$

we have to krige the Hermite polynomials $H_n(Y(x))$ given the $H_n\,[Y(x_a)]$ for the sample points:

$$[H_n(Y(x))]^K = \sum \lambda_a \, H_n[Y(x_a)].$$

Since $Y(x)$ and the $Y(x_a)$ have the same distribution locally, $H_n(Y(x))$ and the $H_n\,[Y(x_a)]$ have the same mean locally. Then it is possible to impose a local unbiasedness condition:

$$\sum_{a} \lambda_{na} = 1.$$

As we saw in Chapter 9 this operation is rather inconvenient because it leads to convergence problems and to a non-optimal estimator. Nevertheless it allows us to lessen the impact of the stationarity hypothesis.

Things are very different when a change of support is involved. Now we want to estimate:

$$f[Y_v] = f_0 + \sum_1 f_n \, H_n(Y_v).$$

This means kriging $H_n(Y_v)$ given the corresponding point support polynomials:

$$[H_n(Y_v)]^K = \sum \lambda_{n\alpha} \, H_n[Y(x_\alpha)].$$

But this time $H_n(Y_v)$ and the $H_n\,[Y(x_\alpha)]$ do not have the same mean locally, for there is no reason for the point support variables $Y(x)$ to have locally the same distribution as the block variable Y_v. Only the raw variables $Z(v)$ and $Z(x_\alpha)$ have the same mean. Imposing an unbiasedness condition $\Sigma_\alpha \lambda_{n\alpha} = 1$ on the kriging system for $H_n(Y_v)$ would be meaningless. Worse, this unjustified action would unfortunately cause some of the estimators to diverge (instead of converging).

So for both gaussian DK and conditional expectation we cannot do without the hypothesis of stationarity when we want to estimate a function of $Z(v)$ such as $I(Z(v) < z)$.

Using uniform conditioning to impose a value

There is one way to overcome this problem of stationarity if we are prepared to group the estimators for all the blocks inside a panel (there must be at least several dozen blocks). This is often the case in mining geostatistics (see Fig 10.3). If the data in the neighbourhood of a particular panel V are too widely spaced, then using methods based on strict stationarity is likely to provide estimates that are too far from the neighbouring values. This happens when the recoverable reserves are being estimated, in particular for the quantity of metal at zero cut-off:

$$\frac{1}{N} \sum_1 Z(v_i) \, I_{Z(v_i) \geq 0}.$$

This is equal to the average grade of all the blocks v_i inside V, that is:

$$\frac{1}{N} \sum_i Z(v_i) = Z(V).$$

In linear geostatistics we normally prefer to use ordinary kriging rather than simple kriging in cases like this. We can wonder whether it would not be possible to base the estimates of the recoverable reserves on the (ordinary) kriged grade of the panel, so that the reserves corresponding to a zero cut-off would equal this value.

Before doing this we must return to the discrete gaussian model. In this, the point values have an anamorphosis function Φ:

$$Z(x) = \sum_0 \phi_n \, H_n[Y(x)]$$

and the block values have a different one Φ_v:

$$Z(v) = \sum_0 \phi_n \, r^n \, H_n[Y_v]$$

where r is the correlation between Y_v and $Y(\underline{x})$ with $\underline{x}$ chosen at random in v. It is determined from the variance of $Z(v)$:

$$\text{var } Z(v) = \sum_1 \phi_n^2 \, r^{2n}.$$

If we make the fairly strong assumption that this model holds for the panel as well as the blocks, then we have:

$$Z(V) = \sum_0 \phi_n \, (r')^n \, H_n[Y_V] \qquad (12.1)$$

where the correlation r' between Y_V and $Y(\underline{x})$ is obtained from:

$$\text{var } Z(V) = \sum_1 (\phi_n)^2 \, (r')^{2n}.$$

So we multiply the point coefficient ϕ_n by $(r')^n$. But this is just the same as multiplying the block coefficients $\phi_n \, r^n$ by $R^n = (r'/r)^n$. The coefficient R which is less than 1 is the correlation between Y_V and $Y_{\underline{v}}$ for a random $\underline{v}$ inside V.

So we change from a block to a panel in the same way as from a point support to a block. Just as we obtained:

$$E[Z(\underline{x})\,|\,Z(v)] = \sum_0 \phi_n \, E[H_n(Y(\underline{x}))\,|\,Y_v]$$

$$= \sum_0 \phi_n \, r^n \, H_n(Y_v) = Z(v)$$

so we can obtain:

$$E[Z(\underline{v})\,|\,Z(V)] = \sum_0 \phi_n \, (r')^n \, E[H_n(Y_{\underline{v}})\,|\,Y_V]$$

$$= \sum_0 \phi_n \, r^n \left(\frac{r'}{r}\right)^n H_n(Y_V) = Z(V).$$

In other words, if we know the panel grade V we also know the mean of the block grades inside the panel. It is exactly the same value. But if we know the panel grade V we can deduce the estimate of any function of $Z(\underline{v})$ (and hence of $Y_{\underline{v}}$) from it. In effect, if we assume, as usual, that the bivariate distribution of Y_V and $Y_{\underline{v}}$ is normal, then we know that given that $Y_V = y_V$, the variable $Y_{\underline{v}}$ has a normal distribution with mean $R \, y_V$ and with variance $1 - R^2$. So we can see that:

$$E[f(Y_{\underline{v}}) | Z(V) = z(V)] = E[f(Y_{\underline{v}}) | Y_V = y_V]$$

$$= \int f(R \, y_V + \sqrt{1 - R^2} u) \, g(u) \, du.$$

In particular we get the following equations for the recoverable reserves. The ore tonnage at a cut-off grade $z = \Phi_v \, (y)$ is:

$$E[I_{Z(\underline{v}) \geq z} | Z(V)] = E[I_{Y_{\underline{v}} \geq y} | Y_V]$$

$$= 1 - G\left(\frac{y - R \, Y_V}{\sqrt{1 - R^2}}\right).$$

The corresponding quantity of metal is:

$$E[Z(\underline{v}) \, I_{Z(\underline{v}) \geq z} | Z(V)] = E[\Phi_v \, (Y_{\underline{v}}) \, I_{Y_{\underline{v}} \geq y} | Y_V].$$

As the block anamorphosis function is determined by its coefficients $\phi_n \, r^n$, we can write, as we did in Chapter 6:

$$\Phi_v \, (Y_{\underline{v}}) \, I_{Y_{\underline{v}} \geq y} = \sum_0 q_n \, H_n(Y_{\underline{v}})$$

with here

$$q_n = \sum_0 \phi_p \, r^p \, U_p^n \, (y).$$

So we obtain the expression for the estimated metal quantity:

$$\sum_0 q_n \, R^n \, H_n[Y_V].$$

To summarize, if we know the panel grade $Z(V)$ then we can deduce the recoverable reserves from it. Then the metal at a zero cut-off grade (i.e. the grade without any selection), $Z(\underline{v})$ will be estimated to be equal to $Z(V)$. In practice we do not know the true value $Z(V)$, so we can think of substituting its estimate $Z(V)^{OK}$ obtained by ordinary kriging. In this type of uniform conditioning, where this value is being imposed on the panel, it is important that Y_V should be determined so as to satisfy the formula corresponding to (12.1):

$$Z(V)^{OK} = \sum_0 \phi_n \, (r')^n \, H_n[Y_V]$$

even if this results in a distribution for the values of Y_V that is not a standard normal distribution. This relation effectively guarantees the underlying quality of the method; that is, it ensures that the estimate of $Z(\underline{v})$:

$$E[Z(\underline{v}) \,|\, Z(V)^{OK}] = \sum_0 \phi_n \, r^n \left(\frac{r'}{r}\right)^n H_n[Y_V]$$

equals $Z(V)^{OK}$.

This method should not be confused with the other types of uniform conditioning described earlier. Compared to them, this one has clear advantages when there are problems with stationarity, even if it is not strictly rigorous.

Changing of support at sample points

In this section we are going to consider a method that is generally referred to as 'service variables with a change of support'. Readers will know that 'service variables' are additive variables that are convenient to use and that can be estimated using linear combinations in order to obtain estimates of other (non-additive) variables. For example, the accumulation and the thickness of a vein can be used in 2-D to estimate the ore grade; similarly the ore tonnage and metal quantity to estimate grade; or the indicator functions for a range of cut-offs in order to estimate the metal quantity and the grade recovered at these cut-offs. We have already seen that indicator cokriging (or disjunctive kriging, as it is also called) can be used in conjunction with non-bias conditions provided that there is no change of support involved.

The idea of this method is simple enough. It is like estimating a variable by using indirect measures. For example, in uranium mines, the U grade of blocks is often estimated from radiometric measurements. To be more precise the U grade

$$U(v) = \frac{1}{v} \int_v U(x)\, dx$$

is estimated from point radiometric measures $Ra(x_\alpha)$. These are transformed into U grades through a regression:

$$E[U(x) \,|\, Ra(x)]$$

which is a function of $Ra(x)$, say $U[Ra(x)]$. Then we assume that $U(x)$ and $U[Ra(x)]$ are equivalent in order to obtain $U(v)$ by estimating

$$\frac{1}{v} \int_v U[Ra(x)]\, dx$$

from the $U[Ra(x_\alpha)]$, for example by kriging. (Of course $U(x)$ and $U[Ra(x)]$ are theoretically not the same, and this creates problems when evaluating the estimation variance.) Setting up the regression and then using it at the sample points presupposes stationarity. But then the kriging can be carried out with a non-bias condition. This avoids the attraction towards the (global) mean that occurs in simple kriging when the sample spacing is too wide.

Now suppose we want to estimate a quantity such as $I(Z(v) \geq z)$ from point samples $Z(x_\alpha)$. For each data point x_α we shall calculate the expected value of $I(Z(v_\alpha) \geq z)$, where v_α denotes the block containing x_α. That is, we shall evaluate:

$$E[I_{Z(v_\alpha) \geq z} | Z(x_\alpha)].$$

Using the gaussian change of support model, this becomes (Exercise 2 (b) in Chapter 10):

$$E[I_{Y_{v_\alpha} \geq y} | Y(x_\alpha)] = 1 - G\left(\frac{y - r\, Y(x_\alpha)}{\sqrt{1 - r^2}}\right).$$

Then we obtain the estimate of $I(Z(v) \geq z)$ that we are looking for by kriging:

$$E[I_{Z(v) \geq z} | Z(x)] = 1 - G\left(\frac{y - r\, Y(x)}{\sqrt{1 - r^2}}\right)$$

where x denotes a point in v, with the point values:

$$E[I_{Z(v_\alpha)\geq z}|Z(x_\alpha)] = 1 - G\left(\frac{y - r\,Y(x_\alpha)}{\sqrt{1-r^2}}\right).$$

This leads us to estimating a variable at one point given the values of the same variable at the sample points. So we can use a non-bias condition in the kriging system.

The problem that arises then is to estimate several indicator functions, or several functions of $Z(v)$, simultaneously. Instead of making a separate kriging for each of them, we can co-estimate them by using (in the gaussian model) the Hermite polynomials. In this case all the functions of $Z(v)$ to be estimated, which are functions of the corresponding gaussian variable Y_v, have to be expanded in terms of Hermite polynomials:

$$f(Y_v) = \sum_0 f_n\, H_n(Y_v).$$

We obtain the required estimates of $E[f(Y_v)\,|\,Y(x)]$ by kriging each of the $E[H_n(Y_v)\,|\,Y(x)] = r^n\; H_n[Y(x)]$ using the values $E[H_n(Y_{v_\alpha})\,|\,Y(x_\alpha)] = r^n\; H_n[Y(x_\alpha)]$. This leads back to kriging the point Hermite polynomials which can be done using a non-bias condition (though this presents the convergence problems mentioned in Chapter 9).

References

This reference list includes a small number of publications which have been chosen because of their theoretical interest or of the quality of the application. Some of them are notes (in French) of the Centre de Géostatistique de Fontainebleau, Ecole des Mines de Paris, which can be obtained by writing to the Librarian at the Centre.

Disjunctive kriging in general, gaussian DK, and the discrete gaussian change of support model

Be careful! Very often Hermite polynomials are not normalized. Hence some modifications are required when expanding the functions.

MATHERON, G. (1973), 'Le krigeage disjonctif', Note of Centre de Géostatistique, 360.

This is the first note on DK. It presents the hermitian (including gaussian), gamma, negative binomial, and Poisson models for DK.

MATHERON, G. (1976), 'A simple substitute for conditional expectation: the disjunctive kriging', *Proceedings, NATO ASI: Advanced Geostatistics in the Mining Industry* (Rome, Oct. 1975), 221–36.

This first publication in English on DK presents the basis of DK and the hermitian (including gaussian) models.

MATHERON, G. (1976), 'Forecasting block grade distribution: the transfer functions', *Proceedings, NATO ASI: Advanced Geostatistics in the Mining Industry* (Rome, Oct. 1975), 237–51.

This paper explains the local estimation of the recoverable reserves ('transfer functions') using DK. The discrete gaussian model and another one (hermitian) are used for the change of support.

MARÉCHAL, A. (1976), 'The practice of transfer functions: numerical methods and their application', *Proceedings, NATO ASI: Advanced Geostatistics in the Mining Industry* (Rome, Oct. 1975), 253–76.

This paper presents the DK of recoverable reserves, its practical problems, and examples from copper and iron deposits.

RENDU, J.-M. (1980), 'Disjunctive kriging: comparison of theory with actual results', *Mathematical Geology*, 12: 305–20.

In this article, DK is used for a direct evaluation of blocks of ore in a gold mine (no change of support is involved). The results are the same as those obtained with lognormal kriging (the grades are lognormally distributed).

MATHERON, G. (1978), 'L'estimation globale des réserves récupérables', Lecture note of Centre de Géostatistique, 75.

This note details the discrete gaussian model and the change of support.

MATHERON, G. (1978), 'Le krigeage disjonctif et le paramétrage local des réserves', Lecture note of Centre de Géostatistique, 76.

This note explains the gaussian DK and its application to the recoverable reserves within the discrete gaussian model. It also includes the estimation of the reserves using the multinormal conditional expectation, and the simplifying uniform conditionings with a gaussian variable.

MATHERON, G. (1979), 'La géostatistique', in *Encyclopedia Universalis*.

These pages give an overall instructive presentation of mining geostatistics: estimation problems encountered, formalism proposed, solutions brought (in situ resources, recoverable reserves, linear and non-linear geostatistics, etc.). They include, in particular, the discrete gaussian model for change of support.

GUIBAL, D., and ZAUPA-REMACRE, A. (1984), 'Local estimation of the recoverable reserves: comparing various methods with the reality on a porphyry copper deposit', *Proceedings, 2nd NATO ASI: Geostatistics for Natural Resources Characterization* (Tahoe, Sept. 1983) (Reidel, Dordrecht).

The different methods, which are reviewed and applied in this paper, are: conditional expectation, uniform conditioning with a gaussian variable, and DK.

ZAUPA-REMACRE, A. (1984), 'L'estimation du récupérable local: le conditionnement uniforme', Thèse de Docteur-Ingénieur, Ecole des Mines de Paris.

The problems of stationarity and the influence of the grid are specifically treated in this work. DK, conditional expectation, uniform conditioning with a normal variable, and uniform conditioning with the kriged panel grade are tested on mining data (gaussian framework).

BOUCHIND'HOMME, J.-F. (1980), 'Estimation de l'uranium récupérable sur les gisements sédimentaires, stratiformes, exploitables à ciel ouvert', Thèse de Docteur-Ingénieur, Institut National polytechnique de Lorraine, Nancy.

In this paper, recoverable reserves are estimated by DK, and also by a kriging of service variables (in the gaussian framework, with and without change of support).

YOUNG, D. S. (1982), 'Development and application of disjunctive kriging model: discrete gaussian model', in T. Johnson and R. Barnes (eds.), *Proceedings, 17th APCOM* (SME/AIME, New York), 544–61.

This paper includes an application of DK to zinc grades in a sulphide deposit.

YATES, S. R., WARRICK, A. W., and MYERS, D. E. (1986), 'Disjunctive kriging: I. Overview of estimation and conditional probability. II. Examples.' *Water Resources Research* 22: 615–21, 623–30.

Part I reviews the theory of gaussian DK. Part II presents an application to mapping electrical conductivity (approximately lognormal), its logarithm, and the probability of exceeding a critical threshold. Note that the block estimates are just regularized point estimates. There is no change of support which makes it possible to estimate the distribution of the block values themselves.

YATES, S. R., and YATES, M. V. (1988), 'Disjunctive kriging as an approach to management decision-making', *Soil Sci. Soc. Am. J.* 52: 1554–8.

Here DK is applied to the contamination of groundwater by viruses: critical concentration in function of the septic tank setback distance.

WOOD, G., OLIVER, M. A., and WEBSTER, R. (1990), 'Estimating soil salinity by disjunctive kriging', *Soil Use and Management*, 6–3: 97–104.

Here DK is used for mapping the salinity of soil and the probability of exceeding critical thresholds from measurements of electrical conductivity.

WEBSTER, R., and OLIVER, M. A. (1989), 'Disjunctive kriging in agriculture', in M. Armstrong (ed.), *Geostatistics, Proceedings, 3rd International Geostatistics Congress* (Avignon, Sept. 1988) (Kluwer, Dordrecht).

Different case-studies are presented in this paper: available phosphorus in the soil, cobalt deficiency, and salinity (electrical conductivity).

WEBSTER, R., and OLIVER, M. A. (1989), 'Optimal interpolation and isarithmic mapping of soil properties: VI. Disjunctive kriging and mapping the conditional probability', *Journal of Soil Science*, 40: 497–512.

This paper is a more developed presentation of the previous paper. The first example concerns pH, exchangeable potassium, and available phosphorus. The second example is on cobalt and copper deficiencies.

LANTUÉJOUL, C., and RIVOIRARD, J. (1984), 'Une méthode de détermination d'anamorphose', Note of Centre de Géostatistique, 916.

This note details the gaussian anamorphosis of an empirical distribution.

RIVOIRARD, J. (1985), 'La convergence des développements en polynômes d'Hermite', *Sciences de la Terre*, Série Informatique 24, Nancy.

In practice, expansions in Hermite polynomials are truncated. The choice of the order of truncation and its consequences are studied in this paper.

Change of support

MATHERON, G. (1981), 'La sélectivité des distributions', Note of Centre de Géostatistique, 686.

This note presents the general properties of the selectivity curves, Cartier's relation, and the influence of support and information effects on the selectivity curves.

MATHERON, G. (1984), 'The selectivity of the distributions and the second principle of geostatistics', *Proceedings, 2nd NATO ASI: Geostatistics for Natural Resources Characterization* (Tahoe, Sept. 1983) (Reidel, Dordrecht).

This paper is a synthesis, in English, of the previous note.

LANTUÉJOUL, C. (1990), 'Cours de sélectivité', Lecture note of Centre de Géostatistique, 140.

Selectivity is presented here as a set of lectures with exercises. The necessary reminders on probability are included. This note is addressed at the reader who wishes to obtain a better theoretical knowledge of the subject.

KAVOURINOS, C. (1987), 'The grade-tonnage curves for a zinc mine in France', in M. Armstrong and G. Matheron (eds.), *Geostatistical Case Studies* (Reidel, Dordrecht).

Support and information effects are illustrated on a selective mining example. The paper include experimental results as well as theoretical ones obtained with the gaussian model.

DEMANGE, C., LAJAUNIE, C., LANTUEJOUL, C., and RIVOIRARD, J. (1987), 'Global recoverable reserves: testing various change of support on uranium data', in M. Armstrong and G. Matheron (eds.), *Geostatistical Case Studies* (Reidel, Dordrecht).

The discrete gaussian, gamma, negative binomial, and mosaic change of support models are used and lead to very similar results for the global reserves.

Isofactorial diffusion models (except the gaussian one)

MATHERON, G. (1984), 'Modèles isofactoriels et changement de support', *Sciences de la Terre*, Série Informatique 18, Nancy.

MATHERON, G. (1984), 'Isofactorial models and change of support', *Proceedings, 2nd NATO ASI: Geostatistics for Natural Resources Characterization* (Tahoe, Sept. 1983) (Reidel, Dordrecht).

This note presents some isofactorial models with their change of support (gamma, negative binomial, Poisson).

HU, L.-Y. (1988), 'Mise en oeuvre du modèle gamma pour l'estimation de distributions spatiales', Thèse de Doctorat de l'Ecole des Mines de Paris.

This explains how to use and fit the gamma model, and describes its advantages over the gaussian model. The gamma model is illustrated through three sets of experimental data.

MATHERON, G. (1984), 'Une méthodologie générale pour les modèles isofactoriels discrets', *Sciences de la Terre*, Série Informatique 21, Nancy, 1–78.

This paper describes the general way to build isofactorial models with a discrete marginal law.

LAJAUNIE, C., and LANTUÉJOUL, C. (1989), 'Setting up the general methodology for discrete isofactorial models', in M. Armstrong (ed.), *Geostatistics, Proceedings, 3rd International Geostatistics Congress* (Avignon, Sept. 1988) (Kluwer, Dordrecht).

This paper explains how to estimate the parameters for the models in the previous note.

RASPA, G., BRUNO, R., and KOKKINIOTIS, P. (1989), 'An application of disjunctive kriging: using the negative binomial model with different change of support models', in M. Armstrong (ed.), *Geostatistics, Proceedings, 3rd International Geostatistics Congress* (Avignon, Sept. 1988) (Kluwer, Dordrecht).

Negative binomial DK with different change of support models is applied to a magnesite orebody.

Indicator kriging and mosaic model

JOURNEL, A. (1982), 'The indicator approach to estimation of spatial distributions', T. B. Johnson and R. J. Barnes (eds.), *Proceedings, 17th APCOM International Symposium* (Society of Mining Engineers of the American Institute of Mining, Metallurgical, and Petroleum Engineers Publ., New York), 793–806.

This is a basic article in favour of the kriging of indicators.

MATHERON, G. (1982), 'La destructuration des hautes teneurs et le krigeage des indicatrices', Note of Centre de Géostatistique, 761.

The author studies when indicator kriging corresponds to indicator cokriging, and introduces the mosaic model.

MATHERON, G. (1984), 'Changement de support en modèle mosaïque'. *Sciences de la Terre*, Série Informatique 20, Nancy.

This paper explains how to make the change of support, when point DK corresponds to kriging the indicators.

Orthogonal indicator residual models

RIVOIRARD, J. (1988), 'Modèles à résidus d'indicatrices autokrigeables', *Sciences de la Terre*, Série Informatique 28, Nancy.

This paper presents the models with orthogonal indicator residuals, including the change of support.

RIVOIRARD, J. (1989), 'Models with orthogonal indicator residuals', in M. Armstrong (ed.), *Geostatistics, Proceedings, 3rd International Geostatistics Congress* (Avignon, Sept. 1988) (Kluwer, Dordrecht), 91–107.

This paper gives a more formal presentation of this model, with the generalization to nested geometrical sets.

Lognormal estimators of in situ reserves

SICHEL, H. S. (1949), 'Mine valuation and maximum likelihood', Master's thesis, University of the Witwatersrand, Johannesburg.

SICHEL, H. S. (1966), 'The estimation of means and associated confidence limits for small samples from lognormal populations', in *Symposium on Mathematical Statistics and Computer Applications in Ore Valuation* (South African Institute of Mining and Metallurgy, Johannesburg), 106–22.

KRIGE, D. G. (1951), 'A statistical approach to some mine valuation and allied problems on the Witwatersrand', Master's thesis, University of the Witwatersrand, Johannesburg.

MATHERON, G. (1962), *Traité de Géostatistique appliquée*: Vol. 1, Mémoires du Bureau de Recherches Géologiques et Minières (Ed. Technip, Paris).

MATHERON, G. (1974), 'Effet proportionnel et lognormalité, ou le retour du serpent de mer', Note of Centre de Géostatistique, 374.

DOWD, P. (1982), 'Lognormal kriging: the general case', *Mathematical Geology*, 14: 474–500.

RIVOIRARD, J. (1990), 'A review of lognormal estimators for *in situ* reserves', *Mathematical Geology,* 22: 213–21.

Evolution of geostatistics

MATHERON, G. (1976), 'Les concepts de base et l'évolution de la géostatistique minière', *Proceedings, NATO ASI: Advanced Geostatistics in the Mining Industry* (Rome, Oct. 1975), 3–10.

MATHERON, G. (1976), 'Le choix des modèles en géostatistique', *Proceedings, NATO ASI: Advanced Geostatistics in the Mining Industry* (Rome, Oct. 1975), 11–27.

MARÉCHAL, A. (1984), 'Recovery estimation: a review of models and methods', *Proceedings, 2nd NATO ASI: Geostatistics for Natural Resources Characterization* (Tahoe, Sept. 1983) (Reidel, Dordrecht), 385–420.

This paper provides a historical and referenced description of the estimation of recoverable reserves from its beginning in the 1970s. The author was himself deeply involved in this effort, with people like G. Matheron, D. Guibal, Y. Touffait, when he was working at the Centre de Géostatistique de Fontainebleau. The paper includes expositions on indicator kriging, disjunctive kriging, conditional distribution, and change of support. Unfortunately the typographic presentation makes it difficult to read.

MATHERON, G., and KLEINGELD, W. J. (1987), 'The evolution of geostatistics', I. C. Lemmer, H. Schaum, and F. A. Camisani-Calzolari (eds.), *Proceedings, 20th APCOM International Symposium* (SAIMM and MINTEK, Johannesburg), 3: 9–12.

KRIGE, D. G., GUARASCIO, M., and CAMISANI-CALZOLARI, F. (1989), 'Early South African geostatistical techniques in today's perspective', in M. Armstrong (ed.), *Geostatistics, Proceedings, 3rd International Geostatistics Congress* (Avignon, Sept. 1988) (Kluwer, Dordrecht), 1–19.

MATHERON, G. (1978), 'Estimer et choisir', *Les Cahiers du Centre de Morphologie Mathématique*, pt. 7 (Ecole des Mines de Paris).

MATHERON, G. (1989), *Estimating and Choosing* (Springer-Verlag, Berlin).

Other references

MATHERON, G. (1965), *Les variables régionalisées et leur estimation*, Thèse de Doctorat d'Etat (Edn. Masson, Paris).

MATHERON, G. (1968), *Osnovy Prikladnoï Geostatistiki* (Mir, Moscow).

MATHERON, G. (1970), 'La théorie des variables régionalisées et ses applications', *Les Cahiers du Centre de Morphologie Mathématique*, pt. 5 (Ecole des Mines de Paris).

MATHERON, G. (1971), 'The theory of regionalized variables and its applications', *Les Cahiers du Centre de Morphologie Mathématique*, pt. 5 (Ecole des Mines de Paris).

DAVID, M. (1977), *Geostatistical Ore Reserve Estimation* (Elsevier, Amsterdam).

JOURNEL, A. (1977), 'Géostatistique Minière', Thèse de Doctorat d'Etat, Fontainebleau.

JOURNEL, A., and HUIJBREGTS, C. (1978), *Mining Geostatistics* (Academic Press, London).

WEBSTER, R., and OLIVER, M. A. (1990), *Statistical Methods in Soil and Land Resource Survey* (Oxford University Press).

PART TWO: CASE-STUDIES

13

Estimating a vein-type deposit using indicator residuals

J. Rivoirard

Introduction

In many vein-type deposits the geometry of the mineralized veins is not known before mining, and its estimation has to be incorporated in the estimation of the reserves. Using a consistent model for the grades and their geometry will lead to more consistent estimates, as we are going to see in an example.

The Lodève uranium deposit is exploited by the Cogéma in the South of France. Geologically it is made up of sedimentary layers, dipping slightly towards the South, and displaced by E–W faulting.

In the part of the deposit we will consider, the uranium is located within several layers, each about 1 metre thick. At the exploration stage, subvertical holes are drilled on a 25 m × 25 m grid. It is always very important for the geologist to build his own interpretation of the deposit. Since the faulting is E–W, the N–S vertical cross-sections provide useful information. They show in particular the role of the major faults (Fig. 13.1).

However, these cross-sections tend to exaggerate the geometrical continuity of the mineralized veins, which has to be reconsidered when drilling new holes. The 25 m grid may seem too large in this case, but indeed there is no hope of determining the geometry of the veins precisely by shortening the grid, as we shall see now.

The upper part of the deposit was exploited in an open-cast mine, with 5 m high levels. Within each level a grid of about 3 m × 3 m blast-holes was made. The radiometry, converted into uranium grade, was measured every 20 cm along each of these 5 m holes. This provides a detailed knowledge of the deposit. Making N–S vertical cross-sections shows how irregular the mineralized veins are, even at this scale (Fig. 13.2).

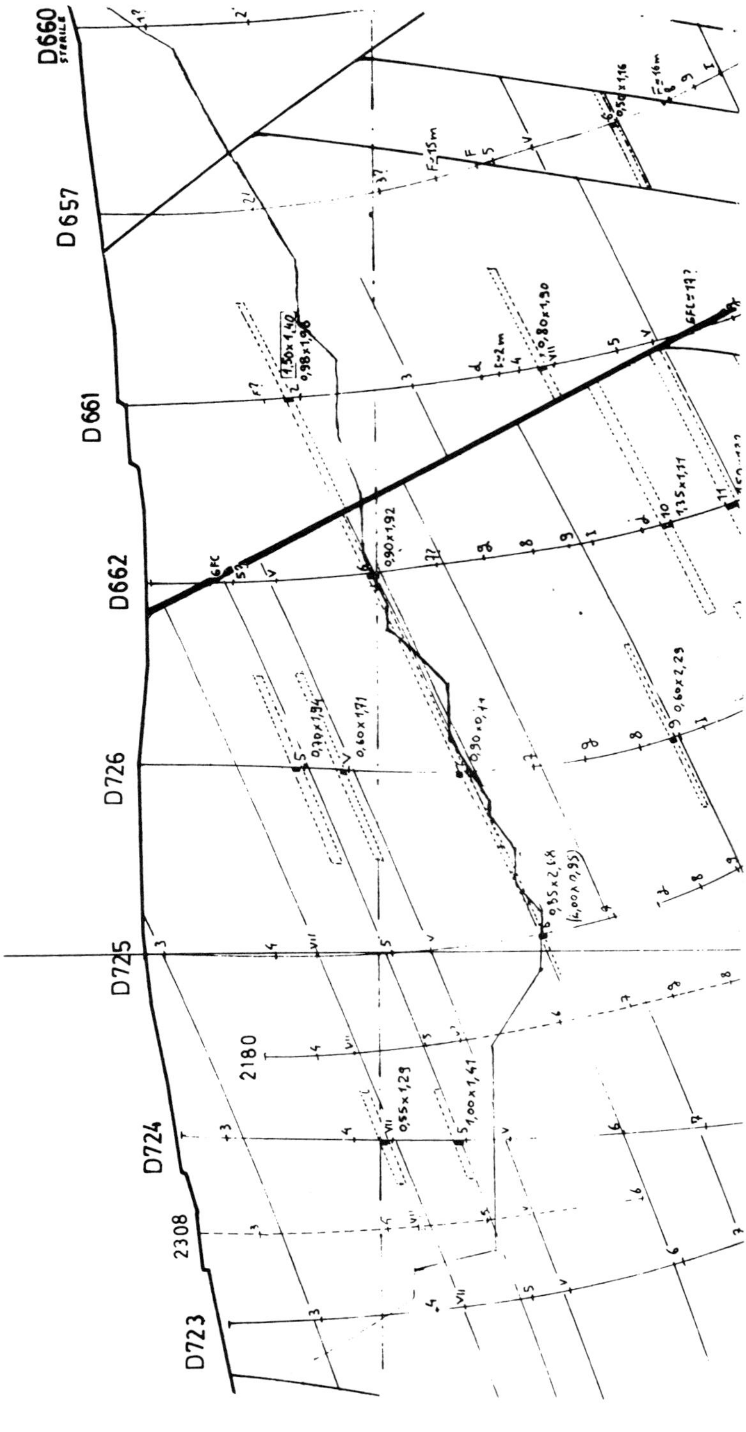

Fig. 13.1 South–North cross-section showing the faults and the geometry of layers, as deduced from the 25 m grid of drill-holes. Working document

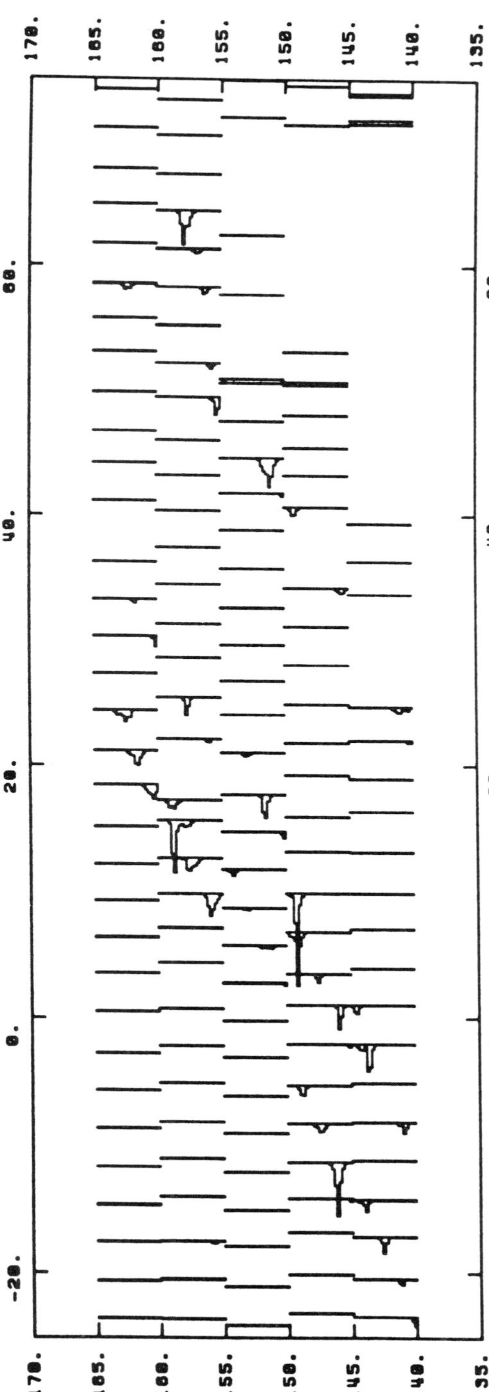

Fig. 13.2 (a) South–North cross-section showing the 3 m grid of blast-holes in the 5 m high benches. The grades plotted along the holes show how irregular the geometry of mineralization is

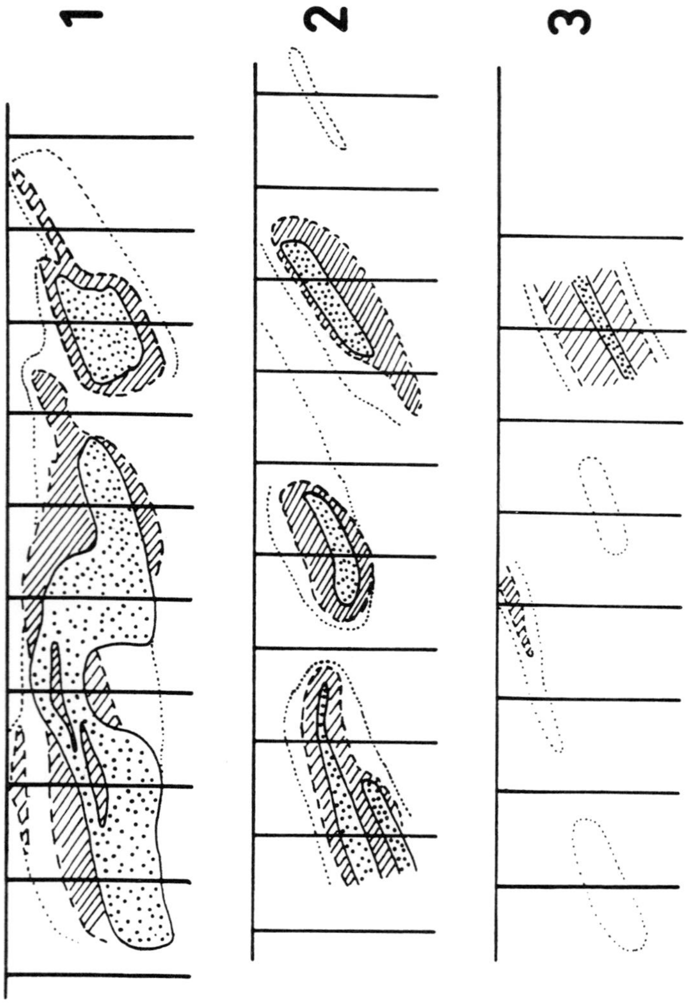

Fig. 13.2 (b) South–North profiles with the blast-holes on 3 m grid. The three profiles are 3 m apart in E–W direction. Geological interpretation showing the continuity of the mineralization. Working document

It is then very convenient to make a geostatistical model of the veins that includes the variability of the geometry as well as that of the grades. This will be done using the indicator residuals.

Variography and modelling

Let $Z(x)$ denote the uranium grade at point x. Outside the mineralized veins this grade is zero. The geometry of the mineralization is then represented by the indicator variable:

$$I_{0.0}(x) = I_{Z(x)>0}.$$

The mean value of this indicator gives the proportion (4%) of the total volume which is mineralized:

$$T(0.0) = P(Z(x) > 0.0) = E[I_{Z(x)>0}] = 0.0409.$$

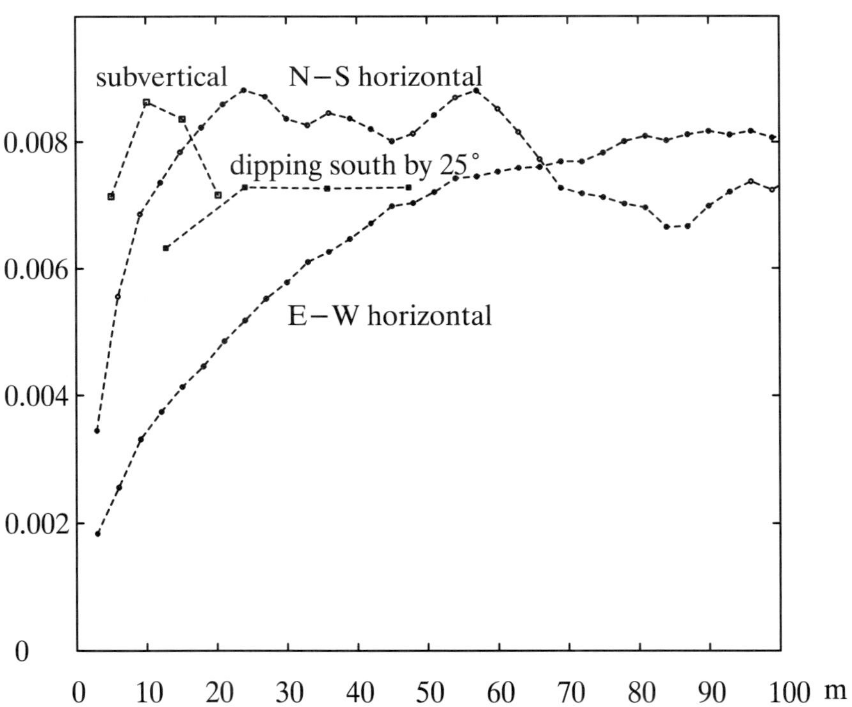

Fig. 13.3 Variograms of the indicator of mineralization (regularized over the 5 m height of levels) in the different directions

Considering the large number of point measurements, it is convenient to regularize this variable over the 5 m of the blast-holes. We then obtain the proportion of ore contained in each hole. (Old geostatisticians will recognize this as a service variable!) Computing the variogram of this variable allows us to quantify the variability or continuity of the geometry (Fig. 13.3).

As expected the best continuity is observed in the E–W horizontal direction. Due to the E–W faulting, this continuity is better than along the direction of the veins dipping south by 25°. The spatial scale of variation in the E–W direction is four times that in the N–S horizontal direction. These differences are summarized into a geometrical anisotropy (coefficients 1.0 in E–W, 2.4 in the direction dipping south by 25°, 10.0 in the third subvertical direction).

The distribution of positive grades is highly skew. Values are expressed as parts per thousand and vary from 0.0 to 39.0, with a mean of 1.24‰ uranium. Table 13.1 gives some basic features. Table 13.2 shows the mean grade within each class of grade.

Just as we have defined the indicator of ore at cut-off zero, we can define the indicators of the mineralization above the cut-off grades 0.5, 1.0, 2.0:

$$I_{0.5}(x) = I_{Z(x)>0.5},\ I_{1.0}(x),\ I_{2.0}(x)$$

as well as their regularized values over the 5-metre holes.

Table 13.1

Values (‰)	Mineralization (%)	% of total
> 0.0	100	T(0.0) = 4.0
> 0.5	57	T(0.5) = 2.3
> 1.0	35	T(1.0) = 1.4
> 2.0	15	T(2.0) = 0.6

Table 13.2

class (‰)	mean grade (‰)
0.0 – 0.5	0.32
0.5 – 1.0	0.71
1.0 – 2.0	1.35
> 2.0	4.41

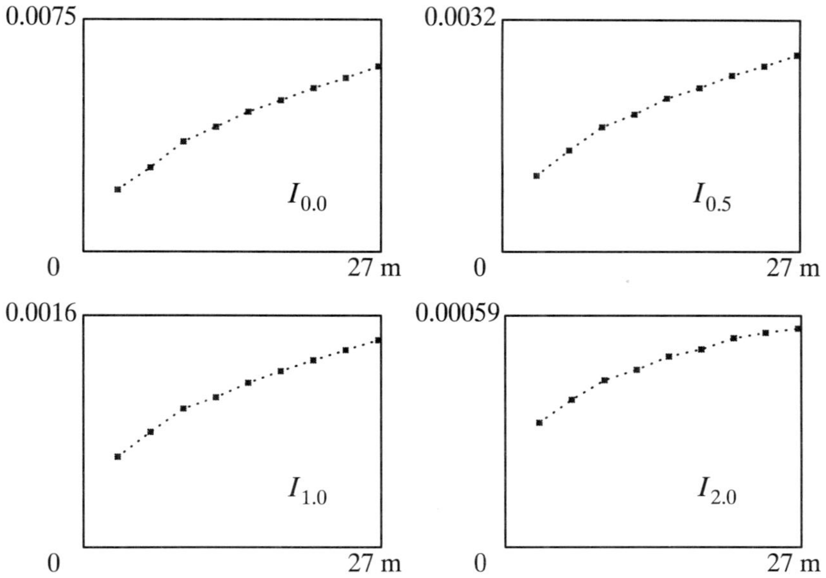

Fig. 13.4 Variograms of the regularized indicators. E–W direction

It is interesting to compare the variograms of the different indicators. In the E–W direction where the structures are best known, we observe a destructuration when the cut-off increases (Fig. 13.4).

In the model with orthogonal indicator residuals, the cross-variogram between two indicators is similar to the variogram of the indicator at the lower cut-off. This relation holds when regularizing the variables, since regularization results in a linear operation on the variograms. In our case it is clear that the cross-variogram of a regularized indicator with a regularized indicator at a higher cut-off is similar to its own variogram (Fig. 13.5).

We define the indicator residuals as follows:

$$R_{0.0}(x) = \frac{I_{0.0}(x)}{T(0.0)} - 1$$

$$R_{0.5}(x) = \frac{I_{0.5}(x)}{T(0.5)} - \frac{I_{0.0}(x)}{T(0.0)}$$

$$R_{1.0}(x) = \frac{I_{1.0}(x)}{T(1.0)} - \frac{I_{0.5}(x)}{T(0.5)}$$

$$R_{2.0}(x) = \frac{I_{2.0}(x)}{T(2.0)} - \frac{I_{1.0}(x)}{T(1.0)}$$

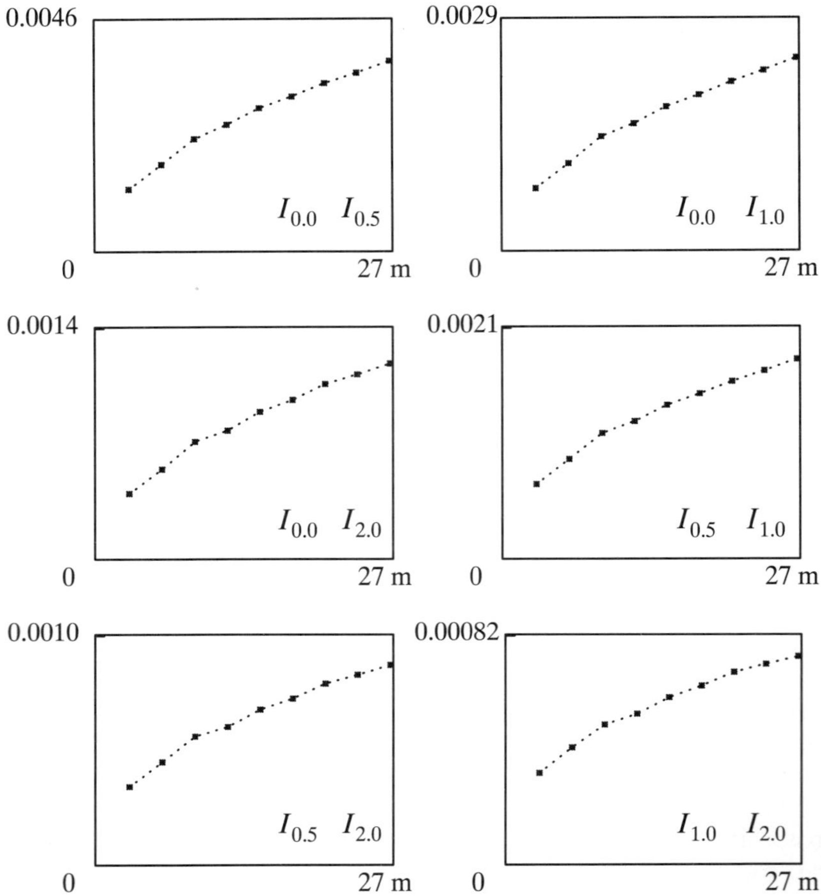

Fig. 13.5 Cross-variograms of the regularized indicators. E–W direction

These residuals have a zero mean. In the model they have no spatial correlation, and kriging each of them gives the cokriging of indicators (i.e. disjunctive kriging). Regularization preserves the non-correlation, and in our case it is more convenient to use the residuals regularized over the 5 m holes. We have just to replace the punctual indicators by the regularized indicators in the formulae.

Note that the first residual corresponds to the geometry of the mineralization. Apart from constant factors, its variogram and its kriging are the same as those of the indicator of ore. As a consequence, the estimate of the ore content, which depends on the positiveness of grades, does not depend on the exact values of these grades. Whether a grade equals 1.0 or 3.0 does not change the estimate of the geometry. This is specific to our model.

Kriging the other residuals requires their variograms. In the best-known direction (E−W horizontal, Fig. 13.6), we observe that these residuals have a nugget component that predominates, and that the range of the remaining structure is about 10 m.

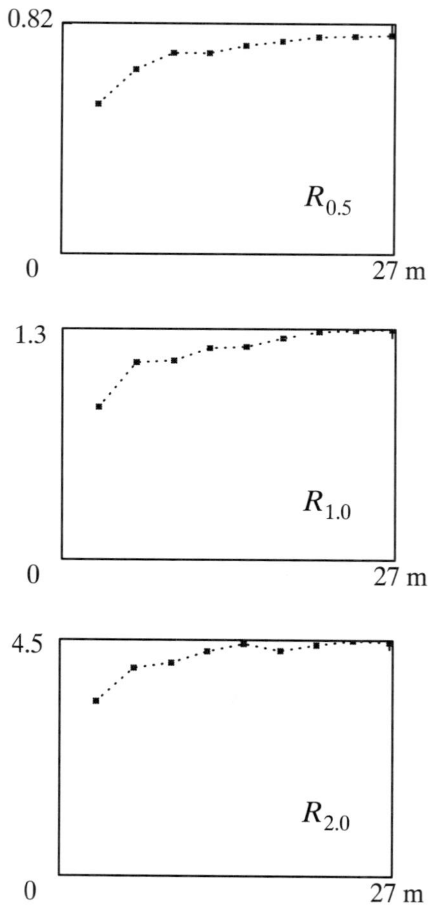

Fig. 13.6 Variograms of the residuals. E−W direction

To summarize these results, the major structure of the deposit arises from the geometry of mineralization (first residual). The distribution of positive grades is itself highly discontinuous, as shown by the other residuals. These two features are present when taking the indicators at different cut-off grades. Their variograms incorporate the structure of the geometry of ore as one part, but the larger the cut-off, the more important is the discontinuous part contributed by the grades.

Estimation of reserves

This model can be used to estimate the ore within each 3 m × 3 m × 5 m block from the grid of blast-holes. But it is not really worth doing this as the results were very close to those obtained directly by kriging. This is because the sampling was so closely spaced.

However the model can be very useful at the exploration stage, when one wants to estimate 25 m × 25 m × 5 m panels V from a 25 m × 25 m grid of measurements. Kriging the different residuals gives the DK proportion of ore above the different cut-offs:

$$[I_{0.0}(V)]^K$$

$$\frac{[I_{0.5}(V)]^{DK}}{T(0.5)} = \frac{[I_{0.0}(V)]^{DK}}{T(0.0)} + [R_{0.5}(V)]^K$$

$$\frac{[I_{1.0}(V)]^{DK}}{T(1.0)} = \frac{[I_{0.5}(V)]^{DK}}{T(0.5)} + [R_{1.0}(V)]^K$$

$$\frac{[I_{2.0}(V)]^{DK}}{T(2.0)} = \frac{[I_{1.0}(V)]^{DK}}{T(1.0)} + [R_{2.0}(V)]^K$$

and the corresponding quantities of metal:

$$[Q_{2.0}(V)]^{DK} = E(Z|Z > 2)\,[I_{2.0}(V)]^{DK} = 4.41\,[I_{2.0}(V)]^{DK}$$

$$\begin{aligned}[Q_{1.0}(V)]^{DK} &= [Q_{2.0}(V)]^{DK}\\ &\quad + E(Z|1 < Z < 2)[I_{1.0}(V) - I_{2.0}(V)]^{DK}\\ &= [Q_{2.0}(V)]^{DK} + 1.35\,[I_{1.0}(V) - I_{2.0}(V)]^{DK}\end{aligned}$$

$$\begin{aligned}[Q_{0.5}(V)]^{DK} &= [Q_{1.0}(V)]^{DK}\\ &\quad + E(Z|0.5 < Z < 1)\,[I_{0.5}(V) - I_{1.0}(V)]^{DK}\\ &= [Q_{1.0}(V)]^{DK} + 0.71\,[I_{0.5}(V) - I_{1.0}(V)]^{DK}\end{aligned}$$

$$[Q_{0.0}(V)]^{DK} = [Q_{0.5}(V)]^{DK}$$

$$+E(Z|0 < Z < 0.5)\,[I_{0.0}\,(V) - I_{0.5}\,(V)]^{DK}$$

$$= [Q_{0.5}\,(V)]^{DK} + 0.32\,[I_{0.0}\,(V) - I_{0.5}\,(V)]^{DK}$$

The weight of the mean in kriging is small (about 15%) for the indicator of geometry, which proves that the kriging gives a really local estimate for this variable. But the structure of the other residuals is very short compared to the grid, and the weight of their means equals 98 or 99%: no really local estimation can be expected, and the kriged residuals are close to their zero mean. We obtain approximately:

$$[I_{0.5}\,(V)]^{DK} = \frac{T(0.5)}{T(0.0)}\,[I_{0.0}\,(V)]^{DK}$$

$$= P(Z > 0.5|Z > 0.0)\quad [I_{0.0}\,(V)]^{DK}$$

$$[I_{1.0}\,(V)]^{DK} = P(Z > 1.0|Z > 0.0)\quad [I_{0.0}\,(V)]^{DK}$$

$$[I_{2.0}\,(V)]^{DK} = P(Z > 2.0|Z > 0.0)\quad [I_{0.0}\,(V)]^{DK}$$

$$[Q_{2.0}\,(V)]^{DK} = \frac{E(Q_{2.0})}{T(0.0)}\,[I_{0.0}\,(V)]^{DK}, \ldots$$

$$[Q_{0.0}\,(V)]^{DK} = \frac{E(Q_{0.0})}{T(0.0)}\,[I_{0.0}\,(V)]^{DK}$$

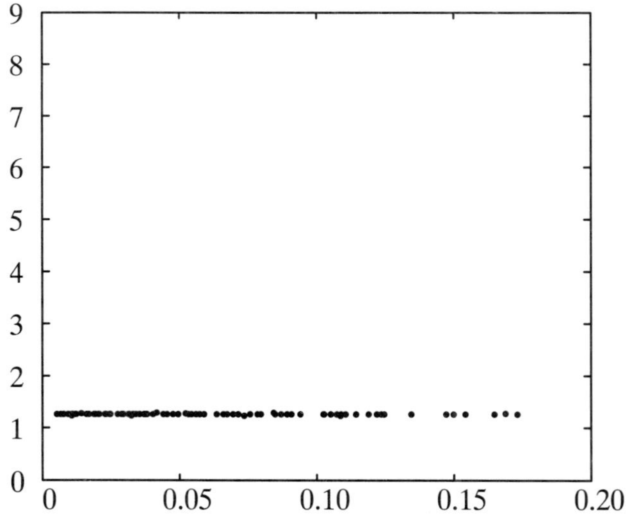

Fig. 13.7 Scatter diagram between grade of ore and ore content, estimated with the indicator residuals

Only the geometry can be estimated really locally. The estimated distribution of grades within the veins is nearly the same throughout the deposit. In particular the metal content $[Q_{0.0}(V)]^{DK}$ is proportional to the ore content $[I_{0.0}(V)]^{DK}$, and the average grade of ore $[Q_{0.0}(V)]^{DK} / [I_{0.0}(V)]^{DK}$ is almost constant everywhere (Fig. 13.7).

Now we have seen how the DK estimates behave, it is instructive to look at how reality does. What we call reality here corresponds to taking all the blast-holes within each panel. The metal content $Q_{0.0}(V)$ increases with the ore content $I_{0.0}(V)$ (Fig. 13.8). The grade of ore $Q_{0.0}(V) / I_{0.0}(V)$ is weakly correlated with the ore quantity but varies a little from one panel to another (Fig. 13.9). The 25 m × 25 m grid does not make it possible to reproduce this.

We might wonder what kriging the indicators instead of cokriging them would give. The structures of the indicators are very different, and the weight of the mean when kriging 25 m × 25 m × 5 m panels with a 25 m × 25 m grid, which is about 15% for the indicator of geometry, equals:

50%	for the indicator at cut-off	0.5
65%		1.0
70%		2.0

It increases with the destructuration.

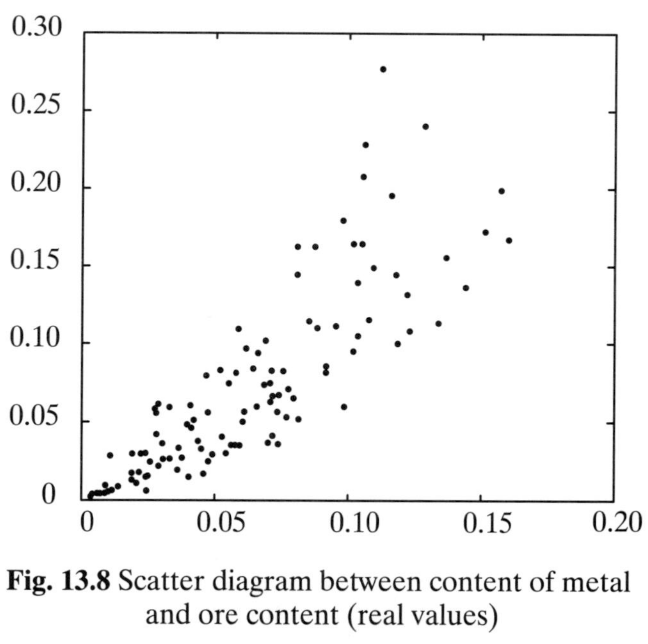

Fig. 13.8 Scatter diagram between content of metal and ore content (real values)

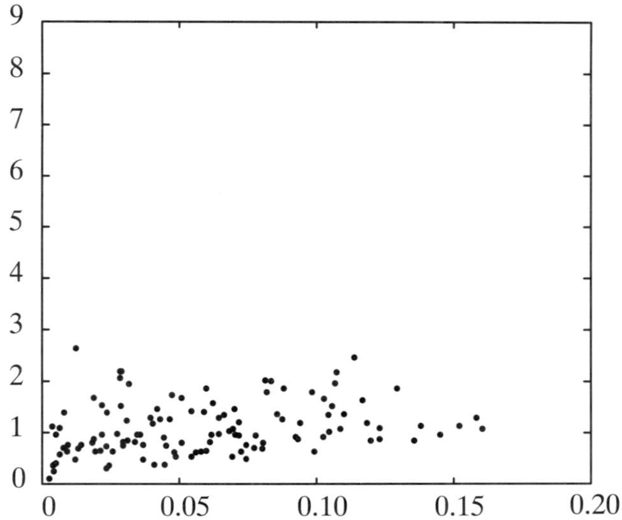

Fig. 13.9 Scatter diagram between grade of ore and ore content (real values)

We know that if we estimate the indicator of geometry using a more variable structure (which is no longer a kriging) we get too smooth an estimate for the ore proportion. This means that locally large concentrations of ore are estimated too small, and conversely the estimation gives too much ore where there is little. Some interesting mining projects (for instance pit contours) could be missed. On the other hand, estimating indicators at high cut-offs with the structure of the geometry would give too much local influence to the large grades. Projects based on the exploitation of such zones may be optimistic.

Let us see what is obtained by kriging the indicators (i.e. each indicator estimated with its own structure) (Fig. 13.10). The smoothing performed when kriging a variable is all the more important as the weight of the mean is larger. As this weight is larger for indicators at high cut-off values than for the geometry indicator, the smoothing is more marked for the proportion of high grades than for the proportion of ore. Zones with a small estimated ore quantity will get a relatively large amount of high grades, and zones with an abundant estimated ore content will have a relatively small proportion of high values. This results in an estimated grade which decreases as the ore quantity increases (even if we mask the smallest values of ore content), which is obviously unsatisfactory (Fig. 13.11).

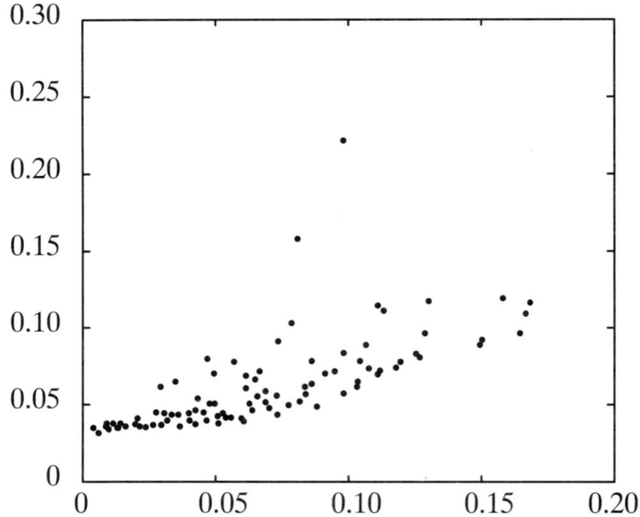

Fig. 13.10 Scatter diagram between metal content and ore content, estimated with the kriged indicators

Remarks

(1) The indicators have been defined on the support of the measurements. The proportions of values we estimate above the different cut-off grades are relative to this almost point support. These geological reserves do not represent a mining selection of larger blocks. A change of support model has to be used for that (Rivoirard 1989).

(2) Using the model with orthogonal indicator residuals requires a discretization of the grades. Here we have considered only the cut-offs 0.0, 0.5, 1.0, 2.0‰ and the grades were then set to the means of classes: 0.32, 0.71, 1.35, 4.41‰. The error of discretization is small in the lower part of the distribution and can easily be made even smaller by taking a few more cut-offs. On the other hand, setting the values ranging from 2.0 to 39.0 to their average value 4.41 may result in a substantial loss of information.
A fine discretization of the tail of the distribution is laborious. A solution consists in adding the orthogonal residual:

$$Q_{2.0}(x) - 4.41\, I_{2.0}(x)$$

which will keep the exact value of the grades when they exceed 2.0 (see Rivoirard 1991). The little improvement we had obtained in our case does not alter our conclusion.

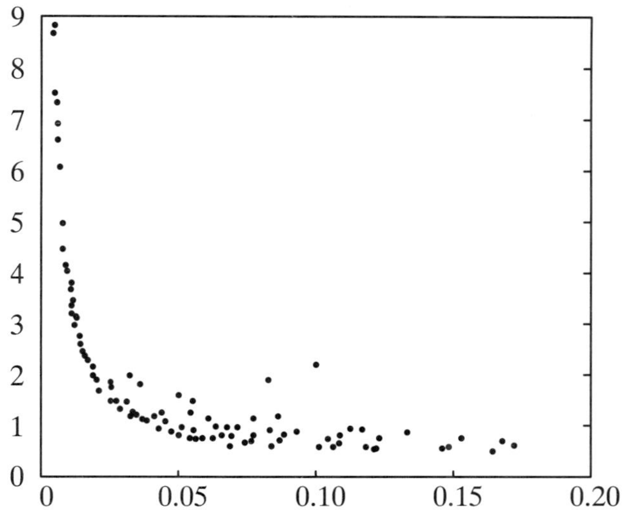

Fig. 13.11 Scatter diagram between grade of ore and ore content, estimated with the kriged indicators

(3) Keeping a single cut-off at 0.0 with the residual between ore and metal:

$$Q_{0.0}(x) - 1.24\ I_{0.0}(x)$$

corresponds to the model used by Bordessoule *et al.* (1989). It is sufficient if one wants only to distinguish waste from mineralization at the zero cut-off.

References

BORDESSOULE, J. L., DEMANGE, C., RIVOIRARD, J. (1989), 'Using an orthogonal residual between ore and metal to estimate *in-situ* uranium resources', in M. Armstrong (ed.), *Geostatistics, Proceedings, 3rd International Geostatistics Congress* (Avignon, Sept. 1988) (Kluwer, Dordrecht), 923–34.

RIVOIRARD, J. (1989), 'Models with orthogonal indicator residuals', in M. Armstrong (ed.), *Geostatistics, Proceedings, 3rd International Geostatistics Congress* (Avignon, Sept. 1988) (Kluwer, Dordrecht), 91–107.

RIVOIRARD, J. (1991), 'Remarques sur les résidus d'indicatrices', *Cahiers de Géostatistique*, pt. 1: 119–35 (Ecole des Mines de Paris).

14

Estimating trace elements in soil: a case-study in cobalt deficiency

Richard Webster

Introduction

All mammals need cobalt, and in the normal way they obtain it in their diets. Grazing livestock obtain their cobalt from the grass and other herbs they eat, which in turn take it up from the soil. Plants do not need cobalt, and so it is possible for plants to grow well on soil that contains none. Animals introduced into such an environment will inevitably sicken and die prematurely for want of the element. Most soil contains some cobalt, but not necessarily enough to support a thriving population of cattle or sheep. Deficiency is serious in some parts of the world.

One region where cobalt deficiency is widespread is south-east Scotland. At one time cattle and sheep suffered severely locally, and deaths were common. This is rare nowadays, but subclinical deficiency is thought still to retard growth and development. To identify causes of sickness and poor growth of livestock, and if possible to forestall them, the local agricultural advisory service, the East of Scotland College of Agriculture, analyses topsoil for farmers and advises corrective treatment if the soil seems deficient. This may involve spraying pasture land with a salt of cobalt, or more usually adding cobalt as a 'salt lick' to the diet of the stock.

The chemistry of cobalt in the soil is complex, and only a small proportion of the element can be taken up by plants at any one time. This is known as 'available' cobalt. It can be extracted from the soil with a dilute solution of some mild acid and measured, and this measurement is the basis for any recommendation. The East of

Scotland College uses 0.5 molar acetic acid for the purpose. Its research has shown that if the available cobalt measured in this way exceeds 0.25 mg/kg of soil (on an air-dry basis) then there should be enough in the herbage for cattle and sheep grazing it. If on the other hand the concentration is less than this value then graziers should be aware, check their animals for deficiency, and take corrective action if necessary. In practice the available cobalt is measured on small samples of soil, and there is the task of using the values obtained to estimate the cobalt concentration for larger areas. In particular the farmer or grazier wants to know whether the soil contains enough cobalt, and this is the problem that we tackle here. The analysis is based on data from an original study by McBratney *et al.* (1982) and developed by Webster and Oliver (1989) and Webster (1991).

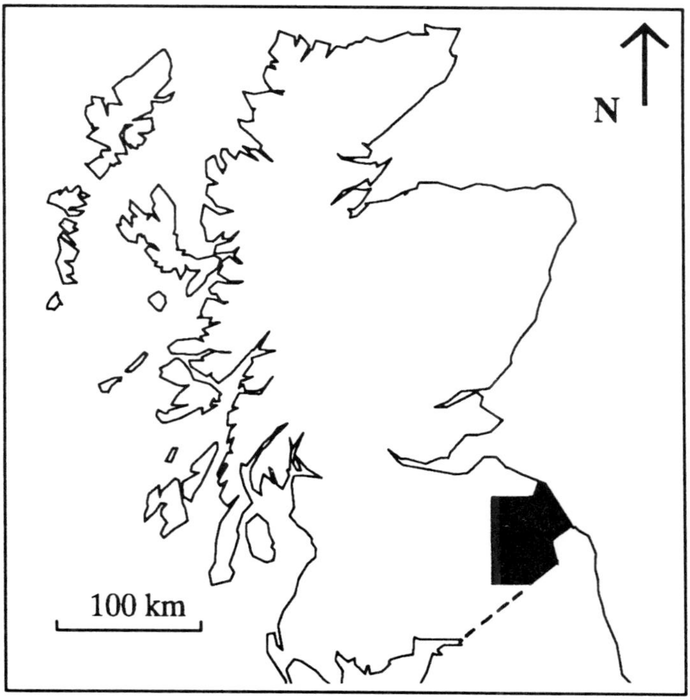

Fig. 14.1 Location of sampled region in south-east Scotland

The survey

Since 1964 the East of Scotland College has analysed topsoil (0–20 cm) from several thousand fields in south-east Scotland. In each field, of area 5 to 10 ha, some twenty cores of soil were taken at random and then bulked and thoroughly mixed for chemical analysis. Larger fields were divided, and each part sampled and analysed separately. The support of the survey is thus the topsoil of approximately 5 to 10 ha.

By 1980 there were records for more than 3500 fields, giving an average intensity of about one sampled field per square kilometre. In the west of the region sampling was very uneven, and not all records were kept. In the east coverage was fairly even and records were complete. This study is therefore restricted to the eastern portion, and specifically to the east of the British National Grid line 360 to 1 km resolution. Fig. 14.1 shows the region, which extends over approximately 1600 km^2, and its position in Scotland. Within it there are 1981 fields for which there are values of available cobalt in the soil and whose grid co-ordinates are known to the nearest 100 m. These constitute the data, and their positions are shown in Fig. 14.2 with symbols whose diameters are proportional to the measured amounts of cobalt.

Although there are fairly many data, each value refers to a support of only 5 to 10 ha, that is a single field. The stock farmer is likely to be more concerned about the average value for his farm, and so the problem is to use the data to decide whether the concentration of cobalt is enough over supports larger than that of the sample, say 1 km × 1 km blocks.

Statistical summary

Table 14.1 summarizes the statistics from the survey. Values of cobalt range from 0.05 mg/kg (very deficient) to 1.0. The mean value is 0.254 mg/kg, which is only slightly more than the critical threshold for deficiency. The data are skewed (the skewness coefficient is 1.6), and in fact the median is less than the threshold. The histogram (Fig. 14.3) displays the spread.

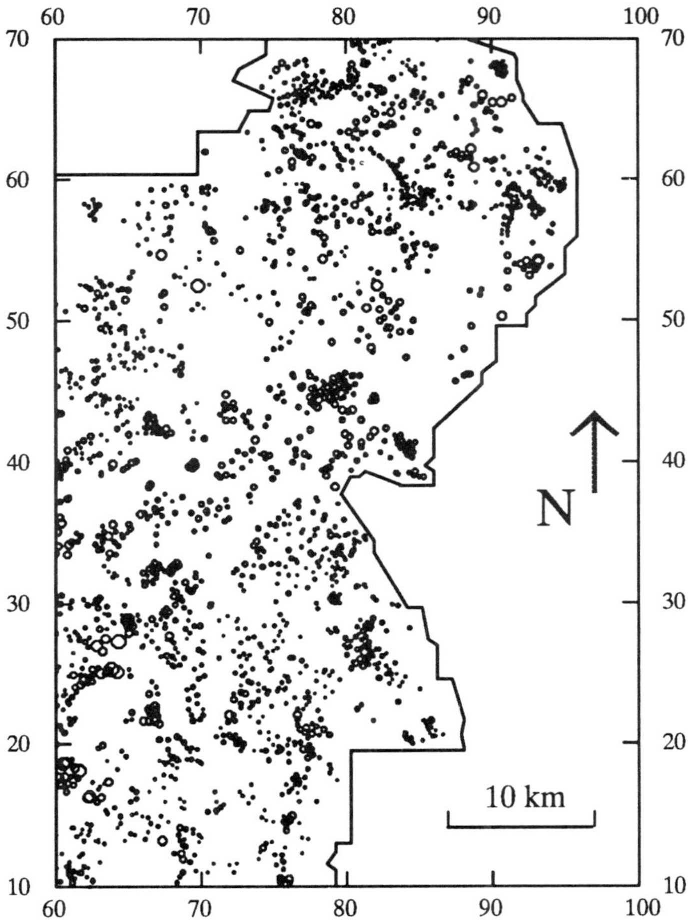

Fig. 14.2 Positions of the fields sampled with symbols of diameter in proportion to the measured concentration of cobalt

Variogram

The variogram has been calculated at intervals of 0.5 km to 25 km using the usual computing formula:

$$\hat{\gamma}(h) = \frac{1}{2n(h)} \sum_{n(h)} [z(x + h) - z(x)]^2$$

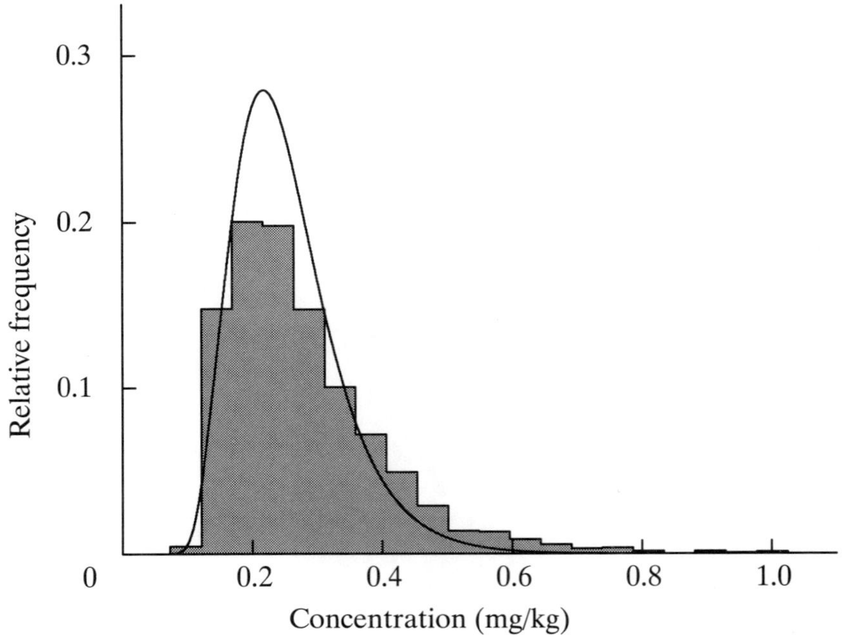

Fig. 14.3 Histogram of cobalt concentration and curve of the block grade distribution

Table 14.1. Summary statistics of available cobalt concentration (original values in mg/kg)

Minimum	0.05
Maximum	1.00
Mean	0.254
Median	0.220
Standard deviation	0.123
Variance	0.0151
Skew	1.61
Kurtosis	4.16
No. of observations	1981

where $z(x)$ are the measured values of cobalt at x, h is the lag distance, $n(h)$ is the number of pairs, $(x, x + h)$, separated by the vector h, approximately, and $\hat{\gamma}(h)$ is the observed semivariance at that lag. The variation is isotropic, and the computed values are averages over all directions. Fig. 14.4 shows the observed values, plotted as points. The continuous curve in the upper part of the figure describes the fitted model. It is double spherical, with formula:

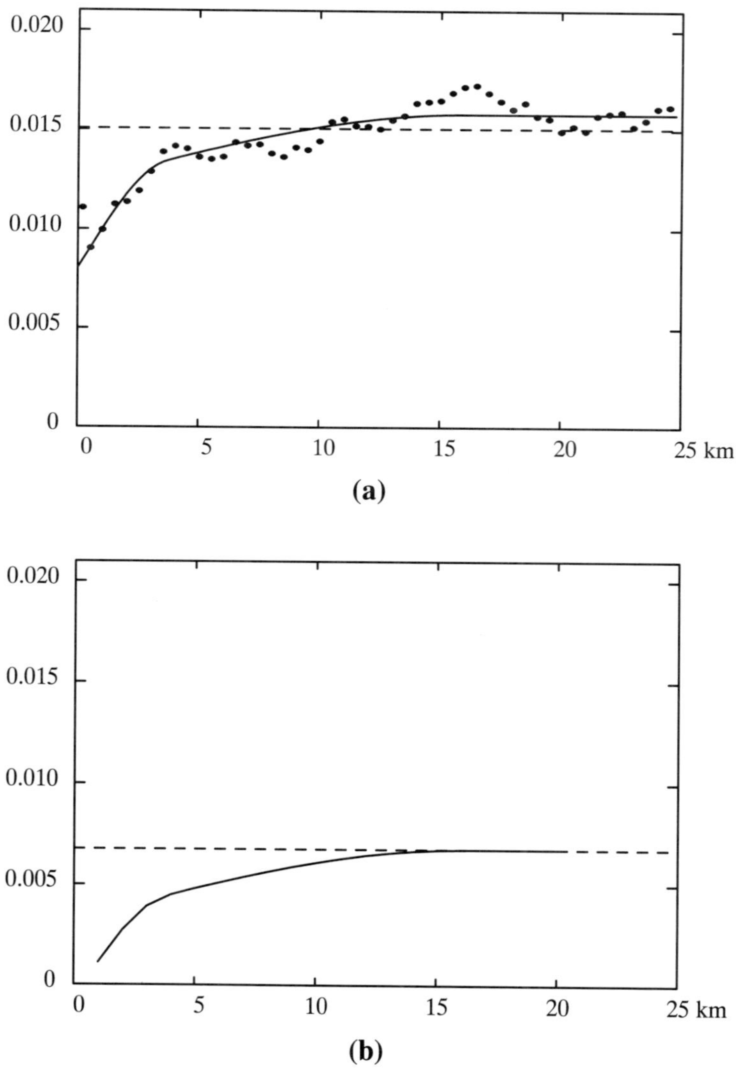

Fig. 14.4 Variograms of (a) the raw data with double spherical model fitted, and (b) the regularized variogram

$$\gamma(h) = c_0 + c_1\left\{\frac{3h}{2a_1} - \frac{1}{2}\left(\frac{h}{a_1}\right)^3\right\} + c_2\left\{\frac{3h}{2a_2} - \frac{1}{2}\left(\frac{h}{a_2}\right)^3\right\}$$
$$\text{for } 0 < h \leq a_1$$

$$\gamma(h) = c_0 + c_1 + c_2\left\{\frac{3h}{2a_2} - \frac{1}{2}\left(\frac{h}{a_2}\right)^3\right\} \quad \text{for } a_1 < h \leq a_2$$

$$\gamma(h) = c_0 + c_1 + c_2 \qquad \text{for } h > a_2$$

$$\gamma(0) = 0$$

The coefficients are given in Table 14.2.

Table 14.2. Coefficients of double spherical models of cobalt

Parameter	Point raw	Block gaussian
c_0	0.008 $(\text{mg/kg})^2$	0
c_1	0.0042 $(\text{mg/kg})^2$	0.44
c_2	0.0036 $(\text{mg/kg})^2$	0.56
a_1	3.7 km	3.7 km
a_2	16.0 km	16.0 km

Ordinary kriging

Using the above variogram and the data we can estimate the amount of cobalt in the soil locally by ordinary kriging. This has been done for 1 km x 1 km blocks using for each block all the data within the block plus those in the eight neighbouring blocks. The average of the kriged estimates is 0.257, which is close to the average of the sample values (0.254), and proves that the samples, though irregularly spaced, were not located preferentially in zones that were either rich or poor in cobalt. The result is mapped in Fig. 14.5. It shows a belt of country through the middle of the region, roughly from south-west to north-east, where the soil is estimated to be rich in cobalt. Elsewhere many of the block estimates are less than the threshold, and farmers whose land falls in these blocks would be advised to check their livestock and perhaps supplement the stock's diet with cobalt.

For completeness we may compute the kriging variance, and this has been done and mapped in Fig. 14.6. In general the variance is fairly small, but at the margins of the region the variance increases sharply because there are few data. There are also gaps in the estimates within the mapped region for want of data locally (see Fig. 14.2 for comparison).

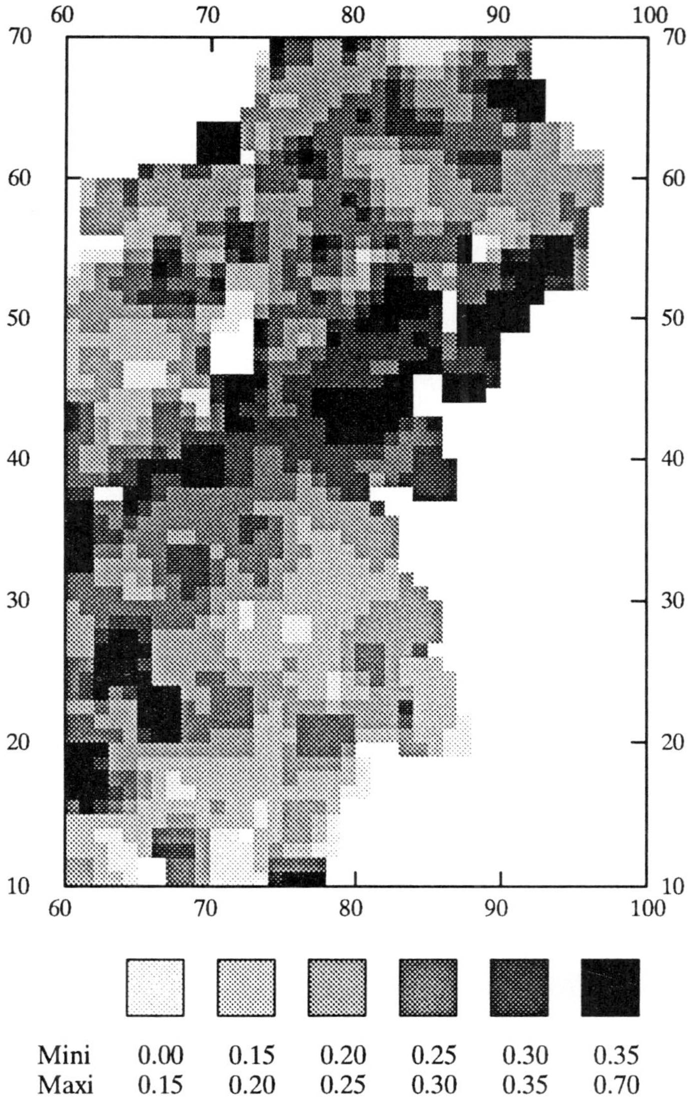

Fig. 14.5 Map of the cobalt concentration estimated by ordinary kriging

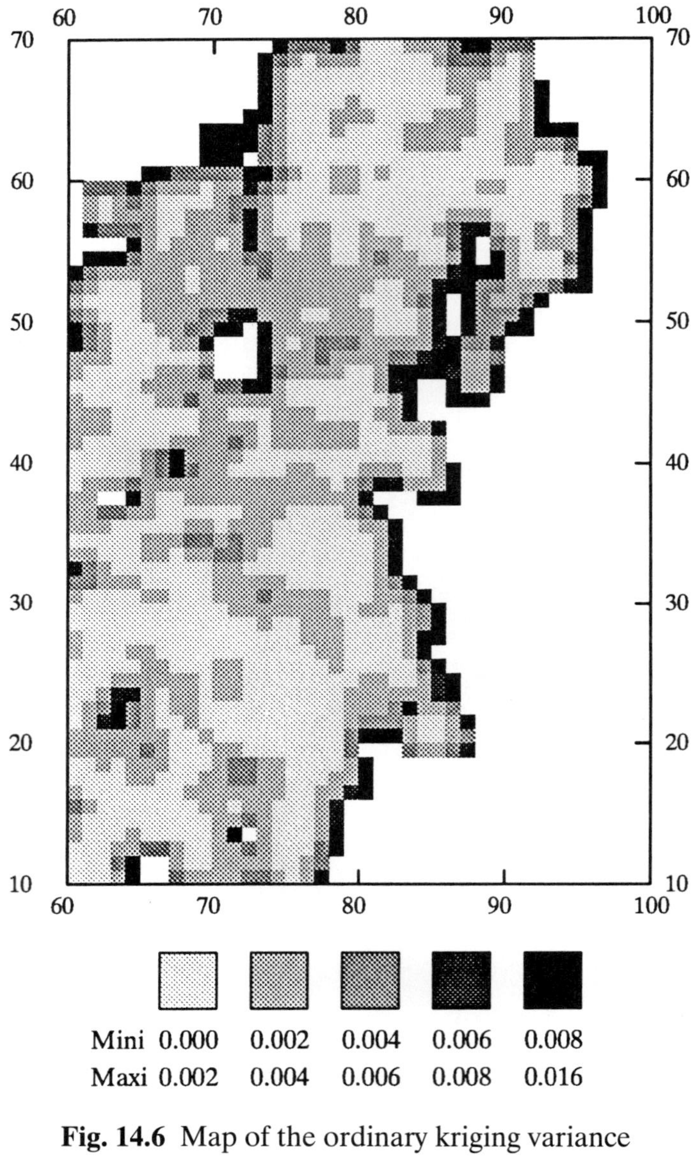

Fig. 14.6 Map of the ordinary kriging variance

Indicator variograms

Our estimates of the cobalt content are not equal to the real (unknown) values, however. So in addition to estimating and mapping the cobalt content of the soil locally we should also like to know the probability that the soil is really deficient over 1 km^2 blocks. This is especially valuable where the estimates are little more than the critical threshold for deficiency. To do this we shall perform a disjunctive kriging to estimate the indicator, $I(Z < 0.25)$, in addition to estimating Z, the cobalt content.

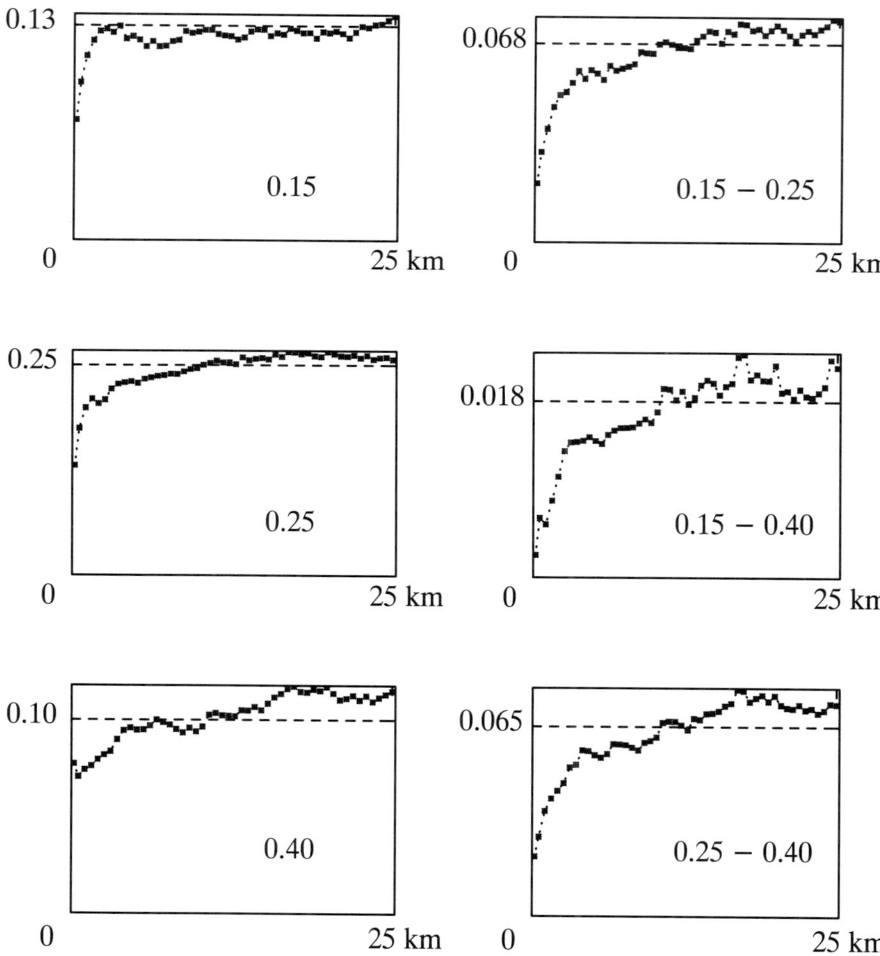

Fig. 14.7 Auto- and cross-variograms of the indicators $I(Z(x) \geq 0.15)$, $I(Z(x) \geq 0.25)$, and $I(Z(x) \geq 0.40)$

To justify our choice of model we compute the cross-variograms of several indicators at cut-off values in the range of the data. We have chosen cut-off values at 0.15, 0.25, and 0.40, which give us four classes containing respectively 12, 40, 42, and 6% of the values. The corresponding indicators are $I(Z(x) \geq 0.15)$, $I(Z(x) \geq 0.25)$, and $I(Z(x) \geq 0.40)$, and their variograms and cross-variograms appear in Fig. 14.7.

We observe in Fig. 14.7 that the cross-variograms are more 'structured' than the auto-variograms. This suggests that we may use a diffusion model for disjunctive kriging, and a gaussian model is then reasonable, as there is no spike in the distribution.

Transformation and change of support

To perform disjunctive kriging on the block support we shall use the discrete gaussian model. Therefore we must transform the data to standard normal form. In this study 30 Hermite polynomials were computed, and Fig. 14.8 shows the transformation that they achieve. The strong concave upward form is another expression of the positive skewness of the raw data.

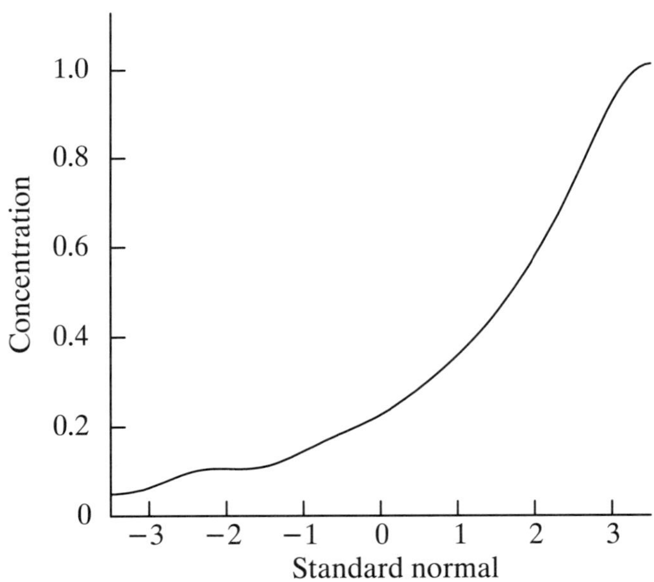

Fig. 14.8 Graph of the transformation of raw concentration to a standard normal variate

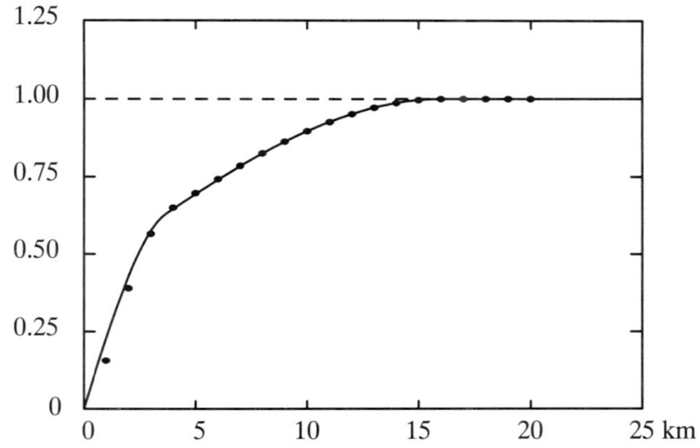

Fig. 14.9 Variogram of the block gaussian variable with estimated values plotted as points and the fitted model drawn as a continuous line

We also wish to change the support from that of the points, in this case fields of area 5 to 10 ha, to 1 km^2 squares. For this we regularize the variogram to this support, with the result shown as the lower curve in Fig. 14.4. Its form is very similar to that of the model fitted to the observed variogram of the raw data. The variances are much less, however, in particular because the nugget variance disappears: it is contained in the within-block variance. The sill variance of the new variogram is 0.00676, compared to 0.0158 for the point variogram, and the change of support coefficient is 0.692. The smaller variance between the blocks shows itself in the distribution function, which is superimposed on the raw histogram in Fig. 14.3. It is more strongly peaked than the original distribution and is lighter in the tails.

The final preliminary to block disjunctive kriging is to compute and model the variogram of the gaussian variable on the block support. This is shown in Fig. 14.9 with the discrete estimates plotted as points, and the fitted model as a continuous line. The model is again double spherical, though without the nugget component, of course. The coefficients are listed in Table 14.2.

Disjunctive kriging

Using the block gaussian variogram and the other relations for the discrete gaussian model the cobalt content and the indicator

$I(Z < 0.25)$ were estimated for the same set of 1 km × 1 km blocks as before. Fig. 14.10 is the result for cobalt content. The pattern is very similar to that obtained by ordinary kriging (Fig. 14.8), as we should expect.

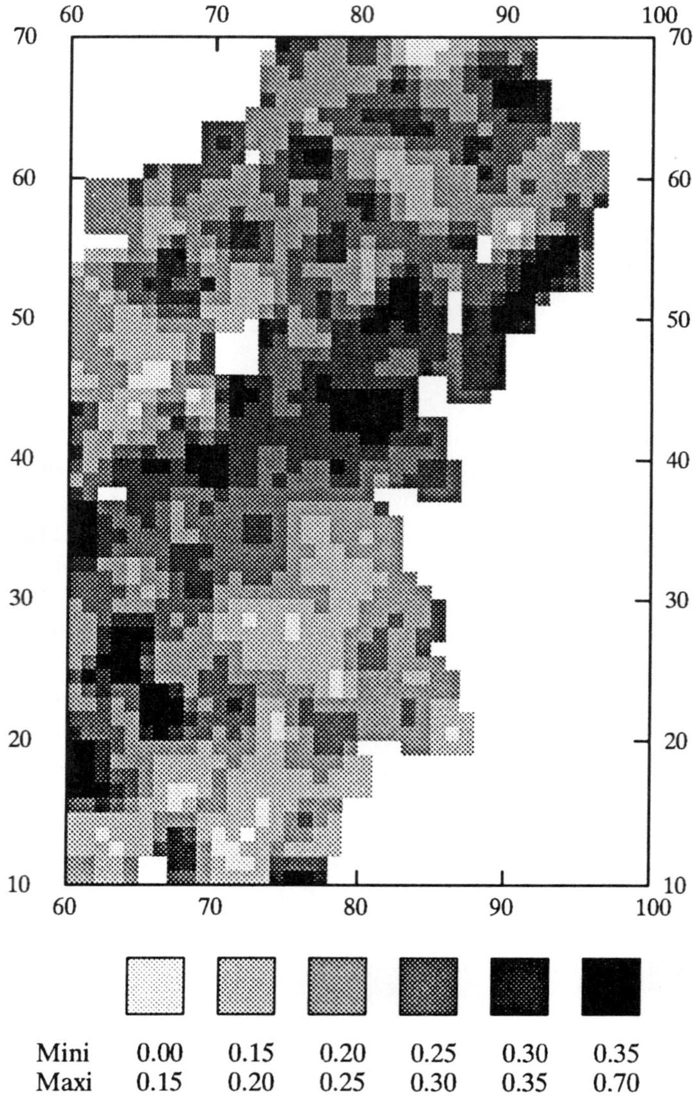

Fig. 14.10 Map of the cobalt concentration estimated by disjunctive kriging

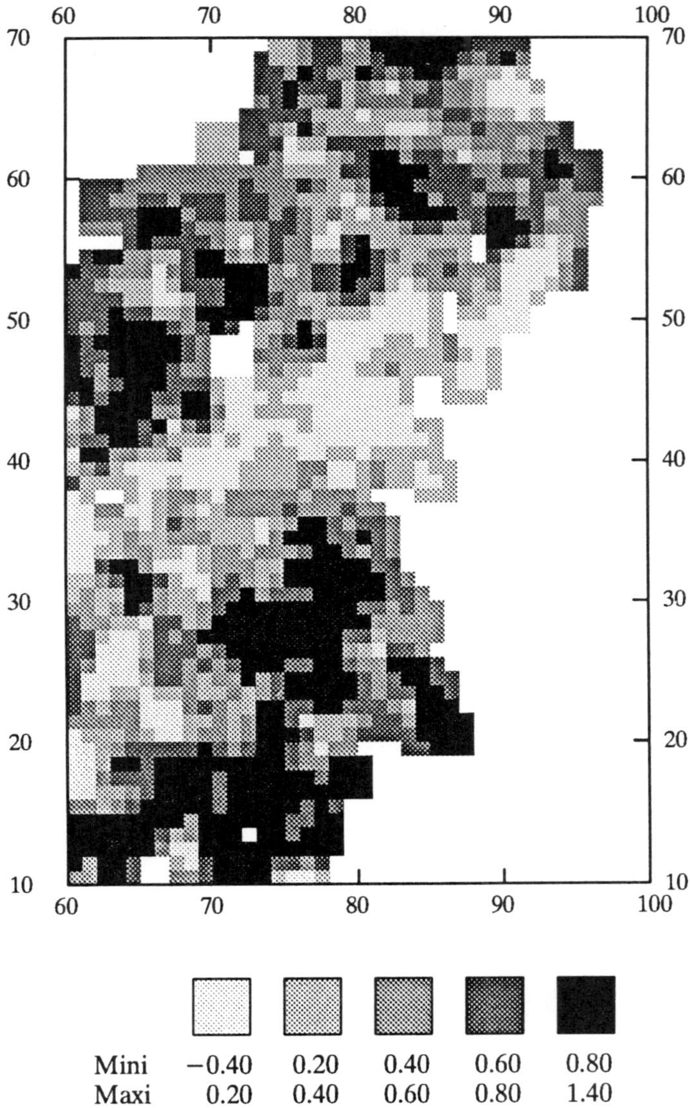

Fig. 14.11 Map of estimates of the indicator $I(Z < 0.25)$

We can also map the estimates of the indicator, the probability that the true value is less than the threshold, and this is done in Fig. 14.11. This appears as the inverse of Fig. 14.10: the smaller the estimate of concentration, the larger is the probability of deficiency. The disjunctive kriging estimates of an indicator are not probabilities in a strict mathematical sense, and so values outside the range [0,1] can be found.

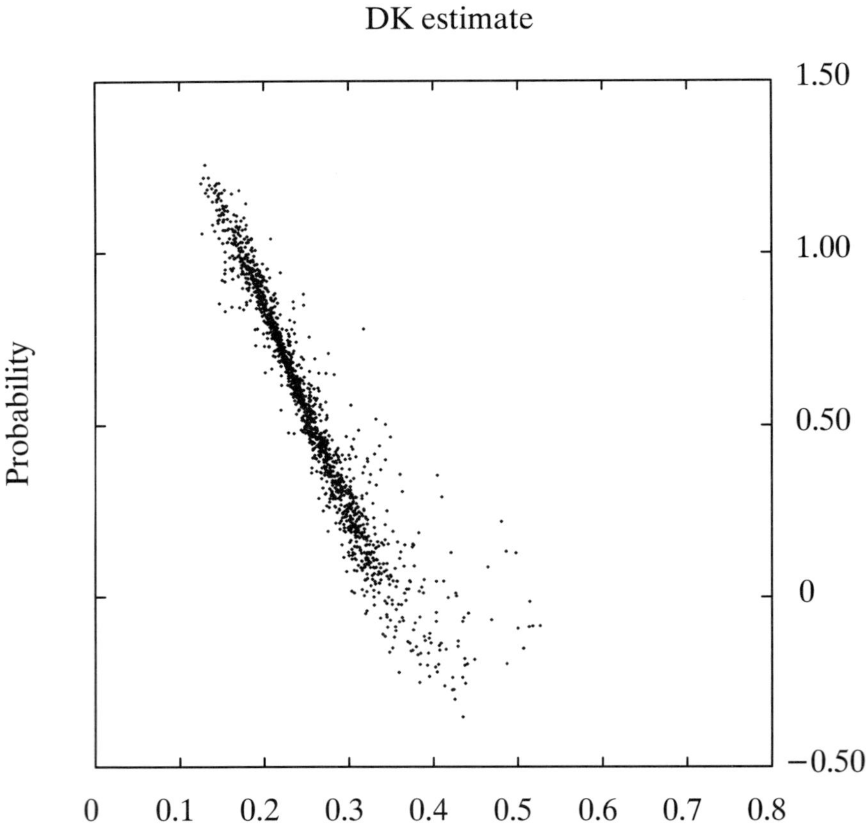

Fig. 14.12 Estimated probability of cobalt deficiency, $I(Z < 0.25)$, against concentration estimated by disjunctive kriging

Relations

Finally in this case-study, we compare some of the results.

First, let us examine the relation between the estimates of the indicator, $I(Z < 0.25)$, and those of concentration (Fig. 14.12). The relation is linear and close with a correlation coefficient of −0.946. To some extent this is to be expected: the two estimates derive from the same procedure. If the estimated concentration is less than 0.25 mg/kg then, as above, the farmer should act to prevent deficiency in his livestock. However, Fig. 14.12 also shows that he will not be assured of enough cobalt in the soil where the estimates lie between 0.25 and 0.30 mg/kg. There is still an appreciable probability that the true value over the 1 km^2 block is less than 0.25 mg/kg.

Let us now look at the relation between the results obtained by ordinary and those from disjunctive kriging. This relation is not quite linear, but the correlation is still fairly strong, $\varrho = 0.944$ (Fig. 14.13). It accords with the similarity of the patterns in the two maps, Figs. 14.10 and 14.11.

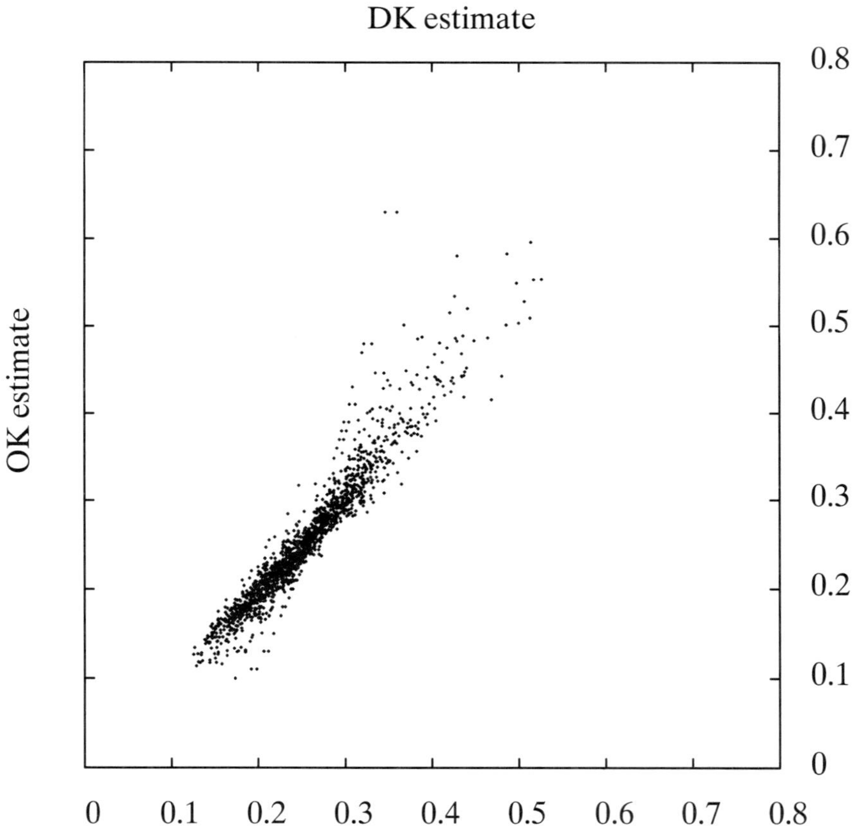

Fig. 14.13 Concentration of cobalt estimated by ordinary kriging against estimate by disjunctive kriging

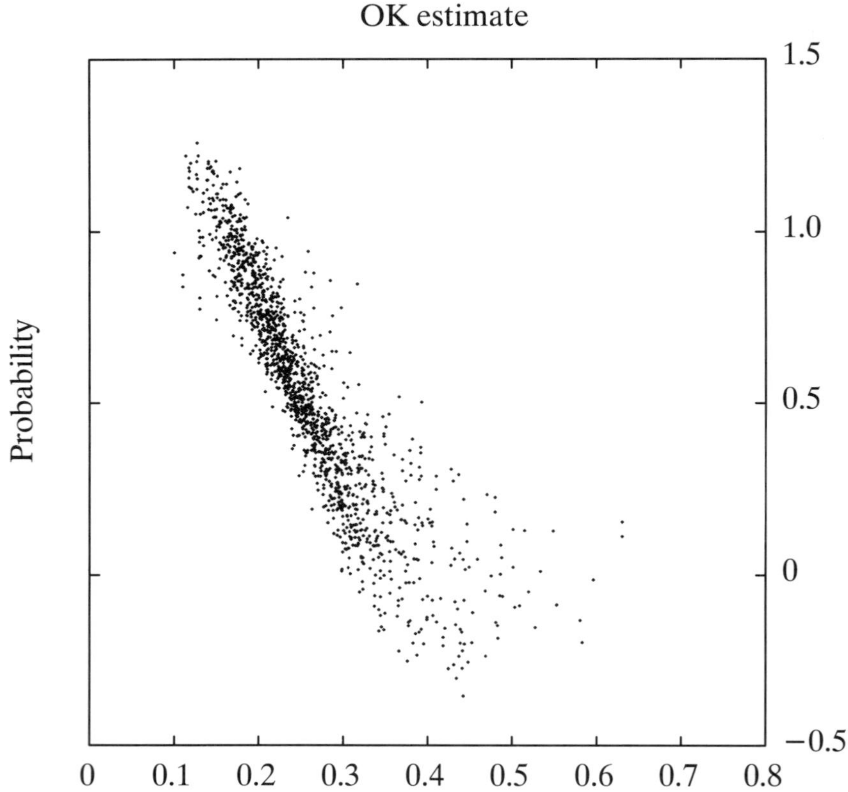

Fig. 14.14 Estimated probability of cobalt deficiency, $I(Z < 0.25)$, against concentration estimated by ordinary kriging

Taking the comparison one stage further, we examine the relation between the ordinary kriged estimates and the estimated probability of sufficiency. The relation is still basically linear, but the correlation is less strong, −0.882 only. In fact we can see from the scatter in Fig. 14.14 the risk that a farmer would run if he relied on ordinary kriging to decide whether to treat his stock against cobalt deficiency. Judging from the results of disjunctive kriging there is a substantial risk of deficiency in the soil up to an estimated value of 0.35 mg/kg.

Conclusion

In this case-study we had measurements of available cobalt in the soil sampled on supports of 5 to 10 ha with a strongly skewed distribution.

We wanted to use them to estimate

(a) the average concentration of cobalt of 1 km × 1 km blocks, and

(b) the probability that the true values in these blocks is less than 0.25 mg/kg, and hence that the soil is deficient in cobalt for grazing livestock.

Disjunctive kriging with change of support enabled us to determine both.

References

MCBRATNEY, A. B., WEBSTER, R., MCLAREN, R. G., SPIERS, R. B. (1982), 'Regional variation in the available copper and cobalt in the topsoil of south-east Scotland', *Agronomie,* 2: 969–82.

WEBSTER, R. (1991), 'Local estimation of soil properties by disjunctive kriging with change of support', *Journal of Soil Science,* 42: 301–18.

WEBSTER, R., OLIVER, M. A. (1989), 'Optimal interpolation and isarithmic mapping of soil properties, VI. Disjunctive kriging and mapping the conditional probability', *Journal of Soil Science,* 40: 497–512.

Acknowledgement

The data were kindly provided by the East of Scotland College of Agriculture, and I thank the College and its staff, Mr H. M. Grey, Dr R. G. McLaren, and Mr R. B. Speirs, for them.

15

Predicting local recoverable reserves for mining panels

Margaret Armstrong

Introduction

Opening up new mines requires enormous capital investments and consequently decision-makers need reliable information on future ore tonnages and metal production. In the case of a selective mining operation they would like ideally to know exactly which blocks are above the cut-off grade. As the initial data from drill-holes is on a wide-spaced grid, it is not possible to pinpoint the locations of blocks above cut-off (using kriging or other estimation methods). However, geostatisticians can provide mine-planners and decision-makers with estimates of the recoverable reserves (that is, the percentage of ore above cut-off and the metal it contains) so that they can carry out the necessary financial analyses. During the feasibility study, the recoverable reserves are calculated first for the whole deposit (i.e. global recoverable reserves) and then if the project goes ahead, for each panel (i.e. locally). Later during mining, much more detailed information becomes available, for example by sampling blast-holes. So at this stage the miner can effectively select (or reject) small blocks (called selective mining units or SMUs).

Several methods could be envisaged for predicting recoverable reserves. Many newcomers to geostatistics are tempted to use kriging. Experienced practitioners know that it is unsuitable for this because it over-smoothes the block grades. More sophisticated techniques based on non-linear estimators or conditional simulation are required.

The objective of this study is to investigate the performance of one particular non-linear technique, disjunctive kriging with the discrete gaussian model, as an estimator of recoverable reserves. In this study we attempt to duplicate the sequence of operations in mining. As a very extensive dataset was available from a precious metal mine, the procedure adopted was to select a subset that can act as 'the initial sample data'. These were then used to predict the recoverable reserves for individual panels. The predicted recoveries (ore tonnage and metal content) could then be compared to actual values obtained from the exhaustive dataset.

The available data

The exhaustive data are available on a 2 m × 2 m grid. Selective mining units, SMUs, will be taken to be this size. The sample data are on a 10 m × 10 m grid and the panels are 20 m × 20 m. Consequently each panel contains 100 small mining blocks and, on average, four samples.

The exhaustive data consist of 5179 point grades. A subset of 300 of these was selected on a regular 10 m × 10 m grid to be used as the 'samples' in the test. Independently of this operation the original values were kriged to obtain the grades of 6242 blocks each 2 m × 2 m. Given the very dense data available for this kriging, these block values could be determined very accurately, and will be considered as the 'real' grades of the blocks. So at this point, on the one hand we have a set of 300 samples that can be used for predicting the recoverable reserves, while on the other hand, we have the corresponding real grades of these blocks. By comparison we can see how well recoverable reserves can be predicted.

Structure of the study

The study is divided into three sections. Section I presents the statistics for the various datasets: the exhaustive set of 5197 data values, the 300 'samples', and the 6242 'real' block grades derived from it. To illustrate the smoothing effect of kriging, the kriged block estimates are obtained from the 300 samples and are compared to the corresponding real ones. Having confirmed that this is not an appropriate way of estimating recoverable reserves we go on, in Section II, to consider the discrete gaussian model and disjunctive kriging. Like many non-linear techniques it is based on a gaussian anamorphosis. After transforming the data to normality we can generate local estimates of the grade/tonnage curves for panel of size 20 m × 20 m. The results are presented and discussed in Section III.

In this study the local reserves have been calculated for a total of seventy-six panels each containing 100 SMUs. Ideally we would like to be able to compare the inverse cumulative histogram which corresponds to the estimated recoverable ore tonnage with the actual recoverable reserves for each of the seventy-six panels, and similarly for the metal contents. While it would be possible to present these 152 comparisons graphically, a more synthetic approach is required if we are to understand what is happening. Consequently after presenting the overall statistics of the estimated recoverable reserves compared to the actual ones, the panels are split into three groups, rich panels, average grade ones, and poor ones, and the statistics are recalculated. Lastly visual comparisons are presented for a row of ten selected panels.

Section I : Use of linear estimation techniques

Original exhaustive data

The original set of data obtained from a mined-out area consists of 5197 Au grades on a regular 2 m × 2 m grid. To preserve their confidentiality the grades were multiplied by a constant in order to have an average of 1.00. After this transformation their basic statistics are as shown in Table 15.1.

Table 15.1. Statistics of the original samples on the 2 m × 2 m grid

No. of samples	5197
Minimum	0
Maximum	13.66
Mean	1.00
Variance	1.206
Coefficient of variation	1.098

These samples lie inside a rectangular region 150 m × 200 m. The experimental variograms were calculated in four directions and were found to be very similar. A double spherical model plus a nugget effect was fitted to the omnidirectional variogram. The parameter values are given in Table 15.2.

Table 15.2. Parameters of variogram fitted to 2 m sample grid data

	Sill	Range
1st spherical model	0.45	9 m
2nd spherical model	0.30	30 m
Nugget effect	0.40	–
Total sill	1.15	–

The total sill of 1.15 is slightly less than the variance (1.206) of the samples. The second structure which accounts for 25% of the variability has a range of 30 m; that is, three times the spacing that will be used later for the reconnaissance grid. This ratio is important in the rest of the study because meaningful spatial predictions and estimates can be made only if the range of the structure is longer than the sample grid spacing.

The 10 m × 10 m grid of samples

A subset of 300 samples on a regular 10 m × 10 m grid was selected from the original data to be used in the same way as would the exploration samples. Their locations are shown in Fig. 15.1. (The empty areas were never mined and so have been excluded from the study.) Its basic statistics are compared to those of the initial 5197 samples in Table 15.3. These confirm that the subset is representative even though it only contains 4% of the initial data. The experimental variograms were calculated in four directions and were found to be very similar to that of the initial data on a 2 m grid.

Table 15.3. Statistics of the subset of 300 samples on a 10 m × 10 m grid and of the original samples on the 2 m × 2 m grid

	Samples on 10 m grid	Samples on 2 m grid
No. of samples	300	5197
Minimum	0.0	0.0
Maximum	7.41	13.66
Mean	0.984	1.00
Variance	1.144	1.206
Coef. of variation	1.086	1.098

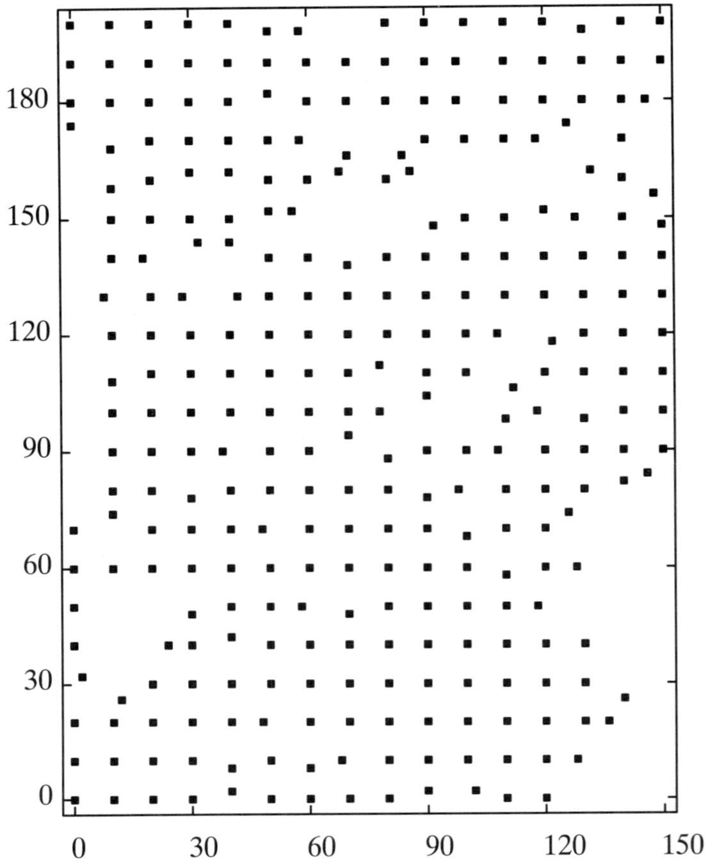

Fig. 15.1 Location of the 300 samples and of the unmined blocks

'Real' grades of the 2 m blocks

All the 2 m × 2 m blocks close to samples were kriged in order to obtain accurate estimates of their grades. A total of 6242 blocks were kriged using the double spherical model given in Table 15.2. A 6 × 6 kriging neighbourhood with a minimum of three data and a maximum of thirty data was used. Table 15.4 gives the statistics of the block estimates.

The kriged block estimates have been obtained from a very closely spaced grid, and consequently can be expected to be very close to the real grades, if these could be measured block by block as the ore is being mined, but this is rarely possible in practice. As these block grades are going to be used in the rest of this study as if they were the real block grades, it is appropriate to ask how good these estimates are. Intuitively given the high ratio of the range to the block size and the one sample to one block ratio we expect excellent estimates, but this should be checked.

Table 15.4. Statistics of the kriged block estimates obtained from the 2 m grid, plus the corresponding kriging standard deviations

	Kriged grades from 2 m grid	Kriging standard deviations
No.	6242	6242
Minimum	0.0026	0.087
Maximum	6.774	0.338
Mean	0.984	0.117
Variance	0.640	0.0027

Some theoretical tests can be applied to see how good the estimates should be. For example, the slope of the linear regression of real grades on the estimated ones can be calculated (under certain assumptions), so can the coefficient of correlation between the real and estimated grades and the weight of the mean in simple kriging (SK). Ideally the regression slope and the correlation coefficient should be 1.0, whilst the weight of the mean in SK should be 0.0. The values for a typical block given in Table 15.5 confirm that the kriged grades obtained from the closely spaced data should be very close to the real but unknown grades. So in the rest of the study these kriged estimates will be considered as being the real block grades.

Table 15.5. Quality tests for kriging a 2 m block from 2 m grid

Regression slope	0.998
Correlation coefficient	0.927
Weight of mean in SK	0.004

Comparison of the kriged block estimates obtained from 10 m grid with real grades

The same set of 6242 blocks was kriged using the 300 samples on the 10 m grid. (To avoid extra complications the variogram from Table 15.2

for the 2 m grid was used.) The basic statistics for these estimates and for the real block grades are given in Table 15.6.

Their means are very similar (0.984 compared to 1.001) and as can be seen from Fig. 15.2, the regression of real grades on the estimated ones ($Z(v)/Z(v)^K$) is linear with slope 0.987 which is very good (except for the last two classes containing very few pairs). However the correlation coefficient is only 0.67 and the dispersions are quite different. Comparison of their histograms shows that the kriged estimates are as expected much smoother and more normally distributed than the real ones. Consequently the recoverable reserves should not be calculated from the kriged block grades.

Having established that kriging is not suitable for estimating the recoverable reserves, even in cases where the kriging can be considered to be good, we now go forward to see whether the more sophisticated non-linear method, DK using the discrete gaussian model, works well in practice in this case.

Table 15.6. Statistics of the kriged block estimates obtained by kriging the samples on the 10 m grid compared to the real block grades

	Real grades	Kriged grades from 10 m grid
No.	6242	6242
Minimum	0.0026	0.0545
Maximum	6.774	5.393
Mean	0.984	1.001
Variance	0.640	0.298

Section II: Use of non-linear estimation techniques

Stationarity

One of the underlying assumptions in non-linear methods is that of stationarity. Although this is difficult to test statistically from a single realization, one can at least check the data for obvious trends. As the data are arranged in a regular grid of 8×10 panels, except for four missing panels in the last column, some simple techniques from exploratory data analysis can be used. Here we plot the average panel grade for each column in the N–S direction (Fig. 15.3). No systematic trends are apparent. Similar results were also obtained in the E–W direction.

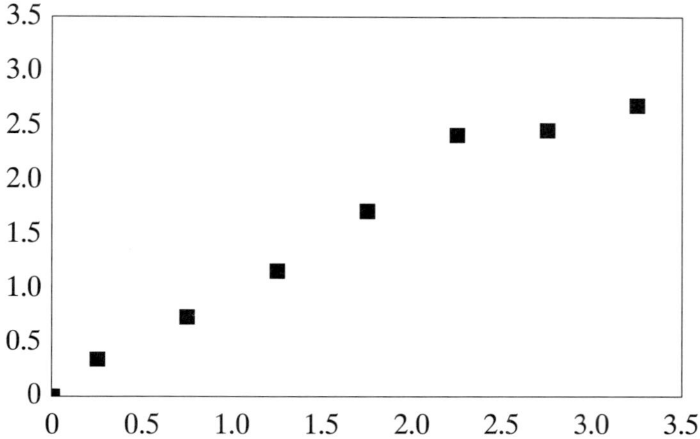

Class	No. points	Mean of $Z(v)$
0.0 – 0.5	975	0.343
0.5 – 1.0	2554	0.734
1.0 – 1.5	1788	1.167
1.5 – 2.0	607	1.721
2.0 – 2.5	208	2.422
2.5 – 3.0	62	2.467
3.0 – 3.5	33	2.696

Fig. 15.2 Mean value of $Z(v)$ per class of values of $Z(v)^K$

Calculation of the gaussian equivalents

As many methods for estimating recoverable reserves including the discrete gaussian model are based on the normal distribution, the first step in the second section of the study is to transform the 300 sample grades to normality using a gaussian anamorphosis. The method is not difficult in itself. If we let z denote the grades and let y be the corresponding gaussian equivalent, then y is chosen so that its cumulative probability distribution $G(y)$ is equal to that of z, called $F(z)$. That is, for each value of z we set y so that

$$G(y) = F(z).$$

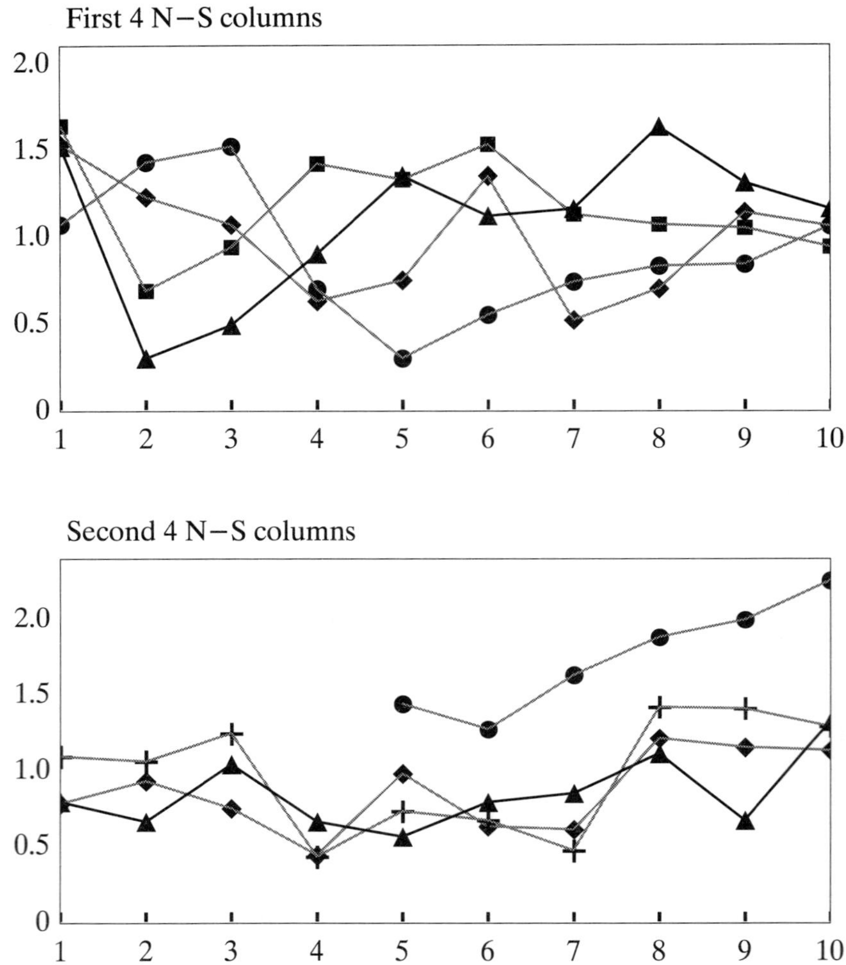

Fig. 15.3 N−S plots of average panel grades to test for non-stationarity

The transformation defined in this way is expressed in terms of a finite number of Hermite polynomials. Typically fifteen to thirty polynomials are enough but it is important to check this. First we test whether the anamorphosis function is monotonic increasing over a large enough range of values of the gaussian equivalent y. In this study the bounds on y were set at -3 and $+3$ but for a simulation of millions or hundreds of thousands of blocks a wider range would have been required. Here we found that at least fifteen Hermite polynomials were needed, and to be on the safe side used thirty of them (Fig. 15.4).

After having checked the quality of the anamorphosis function and fixed the number of Hermite polynomials to be used, the function can then be used to estimate the recovery functions. To show the order of magnitude of the coefficients and the rate of convergence of the anamorphosis, their numerical values and the cumulated variance have been listed for the first twelve coefficients (Fig. 15.4). The cumulated variance should converge to the population variance of 1.1406. In general, the coefficients decrease in absolute value.

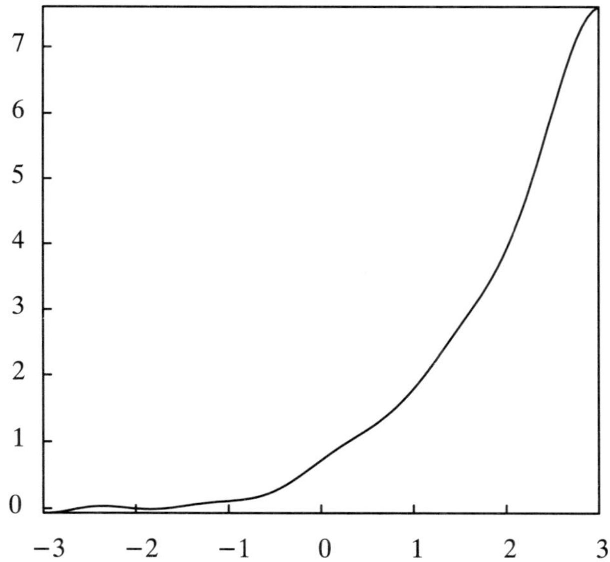

	Hermite coefficient	Cumulated variance		Hermite coefficient	Cumulated variance
ϕ_1	− 0.9498	0.9021	ϕ_7	0.0292	1.1358
ϕ_2	0.4673	1.1205	ϕ_8	0.0259	1.1365
ϕ_3	− 0.1026	1.1310	ϕ_9	− 0.0379	1.1379
ϕ_4	− 0.0073	1.1311	ϕ_{10}	0.0008	1.1379
ϕ_5	0.0321	1.1321	ϕ_{11}	0.0228	1.1384
ϕ_6	− 0.0532	1.1349	ϕ_{12}	− 0.0099	1.1385

Fig. 15.4 Gaussian anamorphosis function

Change of support

The available sample data correspond to a point support, whereas the reserves have to be calculated for mining blocks (which are 2 m × 2 m in this case). We use the discrete gaussian model for the change of support. The first step is to determine the anamorphosis function corresponding to the mining block size. From the user's point of view this comes back to determining the change of support coefficient r which takes values in the range from 0 to 1. The higher the value, the closer the blocks are to a point support. Its value is found iteratively so that the block variance and the point-sample variance satisfy the equation:

$$\text{block variance} = \sum_{n=1} (\phi_n)^2 \; r^{2n}.$$

Here the block variance (obtained by regularization using the fitted variogram for the point support) is 0.65535, compared to the sample variance of 1.1406. The value of r was found to be 0.7923. (Note the block coefficients $\phi_n \, r^n$ converge quite rapidly, since for example r^2 = 0.6277, r^4 = 0.3941, and r^8 = 0.1553.)

Secondly we have to determine the variogram model for the gaussian equivalents corresponding to the block support, from the model for the point support. This is a three-step procedure. First the covariance for the sample support is regularized to give the discretized block covariance corresponding to a 2 m × 2 m support (Fig. 15.5 (a)). The discretized covariance of the corresponding gaussian equivalents is deduced from this (Fig. 15.5 (b)). A variogram/covariance model is then fitted to this. Here it was a double spherical with both sills being 0.5 and ranges of 10 m and 30 m, and no nugget effect. The total sill must be equal to 1.0.

Disjunctive kriging

The basic step in disjunctive kriging is to krige the Hermite polynomials. As there are no correlations to take into account, the polynomials are kriged separately, using simple kriging with a known zero mean. First the kriging neighbourhood is chosen in function of the variogram model. Here it was three panels by three panels, remembering that 20 m × 20 m panels contain four samples each. The user must select the set of cut-off grades for which the recovery functions are to be calculated. As the average grade of the deposit had been set to 1.0, it seemed appropriate to choose the set: 0.0, 0.5, 1.0, 1.5, 2.0. Estimates of the ore and metal recovery were made for each panel for each cut-off grade. The mean recovered grade corresponding to the various cut-offs can then be deduced from these.

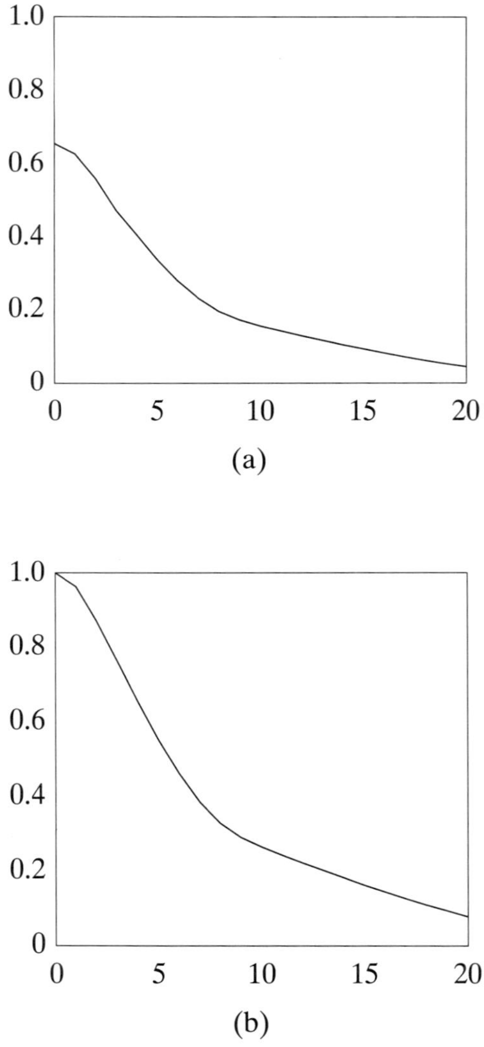

Fig. 15.5 Discretized covariance for (a) block covariance and (b) for the block gaussian equivalents

Section III: Comparison of the DK results with actual grade–tonnage curves

For each panel we now have the recoverable reserves for five cut-offs covering the range of economically interesting grades. The actual grade–tonnage curves calculated from the true grades of the 2 m × 2 m blocks are to be compared with those estimated by DK. Ideally one would like to compare the tonnage and metal predicted by DK with the actual values for all cut-offs and for each of the seventy-six panels but this rapidly becomes fastidious. A more synthetic presentation is required.

First the means of the real and predicted values of the recovered tonnages were calculated for each cut-off (Table 15.7). (The ore tonnage for a cut-off of 0.0 is clearly 100% and so has not been listed.) The metal recovery is also given. The metal content at a 0.0 cut-off is just the average grade. The table prove that overall DK gives accurate predictions. This confirms how well DK works for predicting global reserves.

Table 15.7. Comparison of the averages of the actual and predicted tonnage and metal content for each cut-off

	Ore tonnage		Metal	
cut-off	DK	Real	DK	Real
0.0			0.965	0.967
0.5	0.671	0.661	0.877	0.877
1.0	0.377	0.380	0.661	0.666
1.5	0.195	0.197	0.438	0.447
2.0	0.098	0.100	0.272	0.279

In order to examine the results in more detail, the scatter diagrams were plotted for tonnages and metal content, respectively (Fig. 15.6 and 15.7). When the DK estimate of the panel tonnage closely resembles the real one, the correlation coefficient is near 1.0 as is the slope of the regression. Table 15.8 gives these values for the tonnage and metal recoveries respectively. These values are quite satisfactory.

Table 15.8. Correlation coefficients and regression slopes between the real and DK estimate of tonnage and metal as a function of the cut-off grade

	Ore tonnage		Metal	
cut-off	Correlation	Slope	Correlation	Slope
0.5	0.794	1.30	0.775	1.15
1.0	0.794	1.21	0.751	1.13
1.5	0.681	1.06	0.671	1.06
2.0	0.638	1.09	0.619	1.06

Table 15.9. Comparison between the DK estimated tonnages and the real values for four cut-offs and for sets of high-, medium-, and low-grade panels

	High (%)	Medium (%)	Low (%)
Cut-off 0.5			
Real	87.0	78.5	50.0
DK	87.8	75.5	53.5
Cut-off 1.0			
Real	63.3	47.5	20.2
DK	61.4	45.6	23.1
Cut-off 1.5			
Real	37.3	25.5	7.1
DK	37.9	24.6	8.7
Cut-off 2.0			
Real	19.0	12.8	2.0
DK	22.1	12.8	3.0

Rich, average, and poor panels

It may prove interesting to dissect these results into three classes: poor panels, middling ones, and rich ones. The panels were classified according to their estimated grade, as this is all that users have on hand in practice. Out of a total of seventy-six panels thirty-eight have a DK estimate of average grade (i.e. metal recovery at a zero cut-off) below the population mean of 1.0. Similarly twenty-two panels lie in the range 1.0 to 1.25, while the remaining sixteen lie above 1.25. The average tonnages and metal recoveries above the various cut-off grades were recalculated for these three classes of panels (Tables 15.9 and 15.10). The agreement is excellent for the tonnages figures, and good for the metal recoveries. As the mean recovered grade is just the metal recovered divided by the corresponding tonnage, the DK estimates of this are also good.

Table 15.10. Comparison between the DK estimated metal recoveries and the real values for four cut-offs and for sets of high-, medium-, and low-grade panels

	High	Medium	Low
Cut-off 0.5			
Real	1.32	1.04	0.51
DK	1.40	1.04	0.56
Cut-off 1.0			
Real	1.12	0.81	0.29
DK	1.20	0.81	0.34
Cut-off 1.5			
Real	0.80	0.54	0.13
DK	0.91	0.56	0.17
Cut-off 2.0			
Real	0.53	0.32	0.05
DK	0.64	0.36	0.07

Detailed study of the ten panels

Having established that DK performs well for each class of panels, we now go on to compare the predicted tonnages with the real ones for some of the panels. The area under study is roughly a rectangle eight panels wide in the E–W direction and ten panels long in the N–S direction. Those selected for closer study are the third column (from the left) containing ten panels out of the total seventy-six panels. This column was chosen as its panels all contain more than eighty sub-blocks with known grades (out of a maximum of 100 sub-blocks) and as it goes almost through the centre of the area under study. Consequently the comparisons should be meaningful.

Fig. 15.8 (a) through (j) show the comparison for ore tonnages while Fig. 15.9 (a) through (j) give the comparison for metal. The predicted grade–tonnage curves are good to very good for five panels (3–1, 3–2, 3–6, 3–7, 3–8). The remaining ones are not favourable (3–3, 3–4, 3–5, 3–9, 3–10). Some are overestimated whereas others are underestimated. It is vital to find out why the estimates do not work well in some cases.

To resolve this question, the panels were divided into two classes, according to the absolute difference between the OK estimate of the panel grade and the average of all the sub-blocks in it in the exhaustive dataset (minimum eighty) (Table 15.11). The panels were ranked in order of increasing absolute difference (3–7, 3–8, 3–1, 3–2, 3–6, 3–5 3–9, 3–10, 3–3, and 3–5). The panels with the least absolute difference between the OK panel estimate and the panel average are those where the estimates are good; the others are poorly estimated. Moreover the goodness or badness of the fit is clearly a function of this difference. That is, when OK cannot give a satisfactory estimate of the mean of a panel, DK cannot predict the recoverable reserves, which is hardly surprising.

Table 15.11. Comparison between the OK estimated grades of panels and the averages of all sub-blocks

	OK estimate	Average	Difference	Av. of 4 samples
3–1	1.38	1.55	−0.17	2.04
3–2	1.46	1.29	0.17	0.96
3–3	1.27	1.74	−0.47	1.12
3–4	0.74	0.98	−0.24	0.52
3–5	1.06	0.48	0.58	0.44
3–6	1.48	1.25	0.23	2.70
3–7	0.59	0.56	0.03	0.38
3–8	0.75	0.85	0.10	0.55
3–9	0.96	0.57	0.39	1.09
3–10	0.87	0.40	0.47	1.04

One might ask whether the poorness of the fit is directly related to the numerical values (or the representativeness) of the four samples inside the block. To test this, the average of these four samples was calculated and is shown in the last column of Table 15.11. Looking at panels 3−1 and 3−6, we see an enormous difference between the average of the closest four samples and true panel average, and yet these panels were among those where both OK and DK gave satisfactory or good results. Clearly some other samples in the OK and DK kriging neighbourhoods pulled these estimates into line. So the accuracy of the recoverable reserves prediction does not depend just on the samples in the panel under study.

Having seen the comparisons between the predicted recoverable reserves and their true values, it is also interesting to see how much difference there is between the various grade−tonnage curves. Do they all look alike or are there marked differences between them? One might wonder whether DK produces 'look-alike' estimates for all panels. Fig. 15.10 (a) and (b) present the ore tonnage as a function of cut-off grade for panels 3−1 to 3−5, and for 3−6 to 3−10 respectively. (When figures were superimposed on one graphic it was unreadable.) Similarly for metal in Fig. 15.11 (a) and (b). There are significant differences between these curves. Moreover the poorly predicted panels are not systematically highest or lowest.

Conclusions

This case-study presents a detailed comparison between the local estimates of recoverable reserves obtained using disjunctive kriging with the discrete gaussian model and the true recoveries (ore tonnages and metal quantities) for a set of seventy-six mining panels. The true values were known because a large area in a gold deposit had been mined out exhaustively.

From the original set of 5197 values, we were able to (1) extract a subset of 300 'samples' on a regular 10 m × 10 m grid which were used as input data for the DK estimates, and (2) independently of this operation, obtain the 'real' grades of each small block of size 2 m × 2 m by ordinary kriging. Each of the large panels contains 100 small mining blocks and, on average, four samples.

This study is divided into three sections. Section I of the report presents the statistics for the various datasets: the exhaustive set of 5197 data values, the 300 'samples', and the 6242 'real' block grades derived from it. To illustrate the smoothing effect of kriging, the kriged block estimates are obtained from the 300 samples and are compared to the corresponding real ones. Their means are very similar and the slopes of the linear regression of real grades on the estimated ones are near 1.00. However the correlation coefficient for the kriging using the 300 samples is only 0.67 and the dispersions are quite different. Consequently the recoverable reserves should not be calculated from the kriged block grades.

Having confirmed that this is not an appropriate way of estimating recoverable reserves we go on, in Section II, to consider the discrete gaussian model and disjunctive kriging. Like many non-linear techniques it is based on a gaussian anamorphosis. After transforming the data to normality and checking the quality of this transformation, we can estimate the grade–tonnage curves for the seventy-six panels of size 20 m × 20 m. The results are presented and discussed in Section III.

Ideally we would like to be able to compare the inverse cumulative histogram which corresponds to the estimated recoverable ore tonnage with the actual recoverable reserves for each of the panels, and similarly for the metal contents, but this would rapidly become fastidious. First to summarize the results, the overall statistics of the estimated recoverable reserves were compared to the actual ones. The results are very good. Next, the panels were split into three groups, panels estimated rich, average, or poor, and the statistics were recalculated. No systematic over- or under-estimation was found. Lastly detailed comparisons were presented for a row of ten selected panels. The predicted grade–tonnage curves are good to very good for five of the ten panels. Some of the poorly estimated ones were overestimated whereas others are underestimated.

When the panels had been ranked in order of increasing absolute difference between the OK estimate of the panel grade and its true value, it was obvious that the goodness or badness of the fit depended on this difference. That is, when OK cannot give a satisfactory estimate of the mean of a panel, DK cannot predict the recoverable reserves, which is hardly surprising.

Surprisingly, the poorness of the fit is not directly related to the numerical values (or the representativeness) of the four samples inside the block. There are panels where there was an enormous difference between the average of closest four samples, and yet these panels were among those where both OK and DK gave satisfactory or good results. Clearly some other samples in the OK and DK kriging neighbourhoods pulled these estimates into line. So the accuracy of the recoverable reserves prediction does not depend just on the samples in the panel under study.

This study shows that disjunctive kriging using the discrete gaussian model gives good estimates of the local recoverable reserves, whenever ordinary kriging gives good estimates of the average grade of the panel. It would be difficult to imagine getting better estimates for recoverable reserves than for the panel grade.

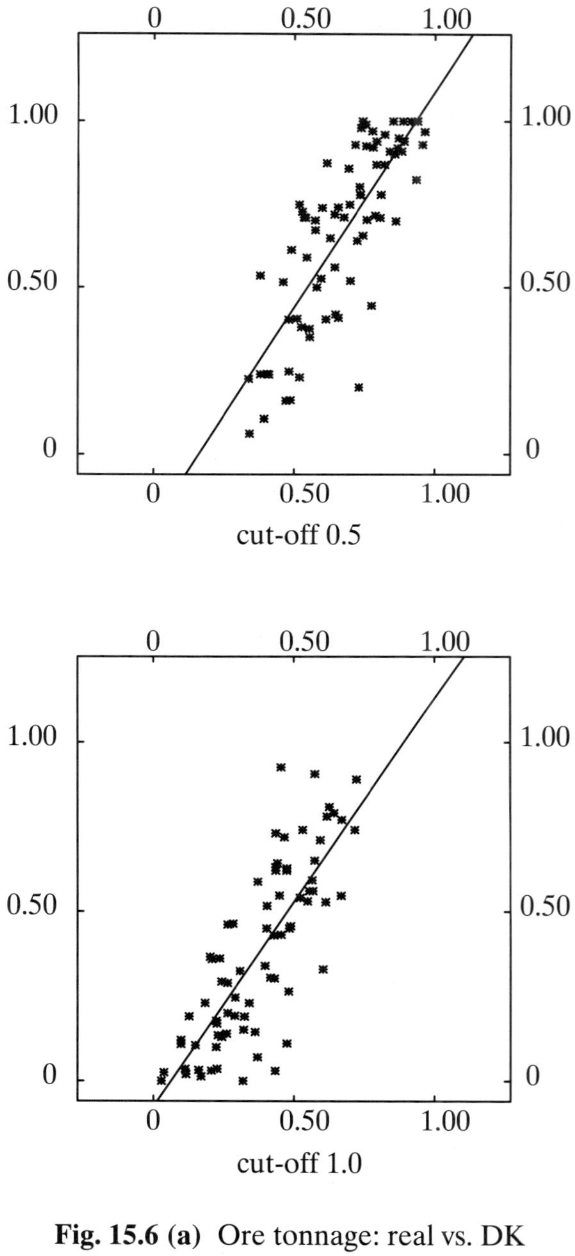

Fig. 15.6 (a) Ore tonnage: real vs. DK

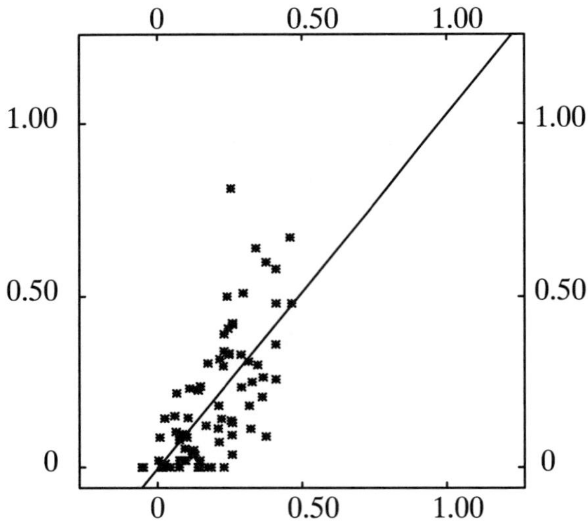

cut-off 1.5

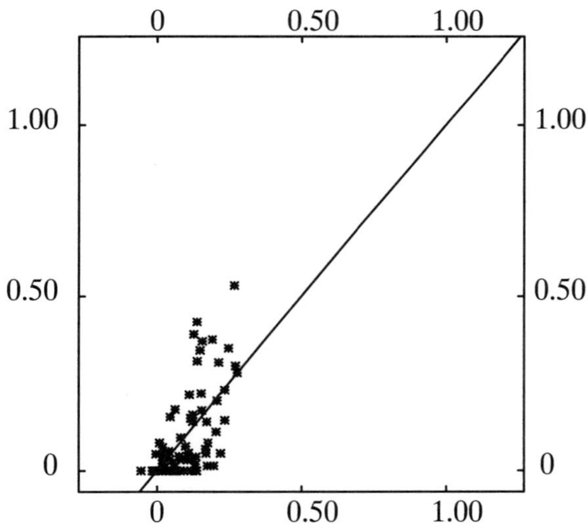

cut-off 2.0

Fig. 15.6 (b) Ore tonnage: real vs. DK

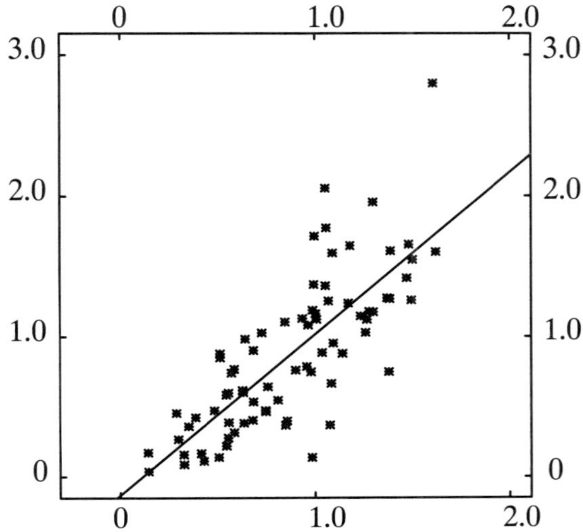

cut-off 0.5

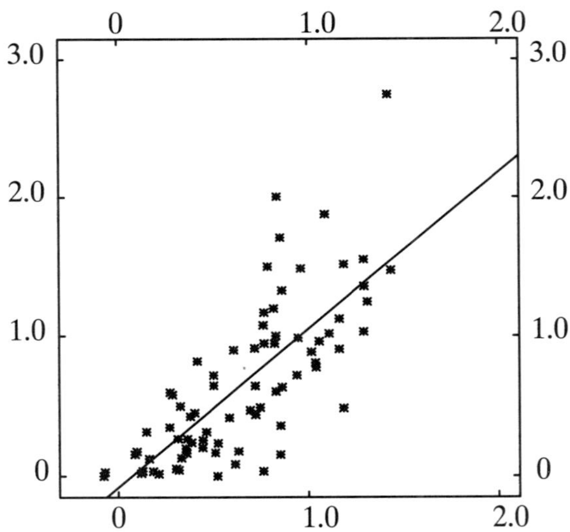

cut-off 1.0

Fig. 15.7 (a) Metal: real vs. DK

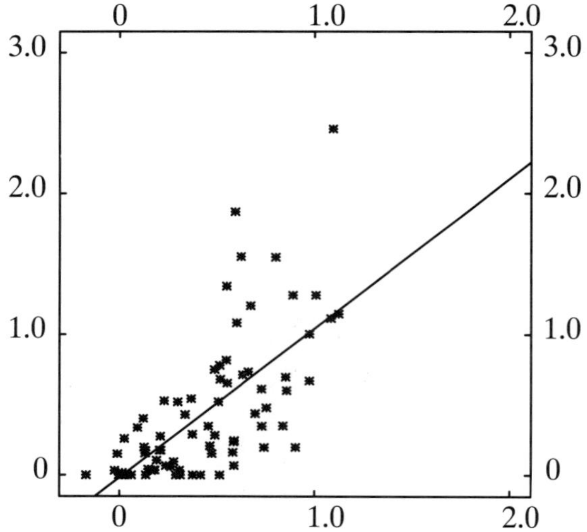

cut-off 1.5

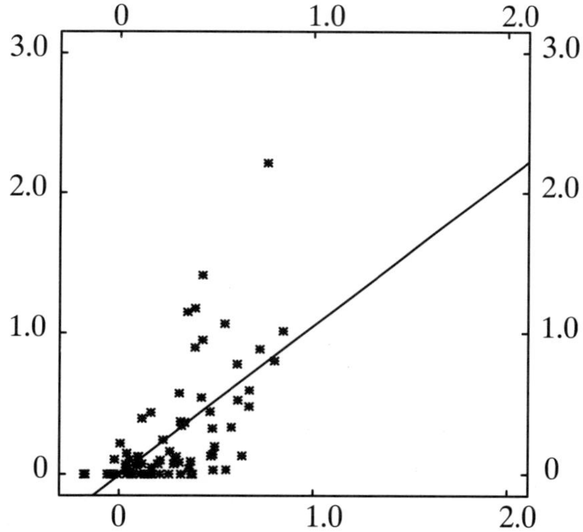

cut-off 2.0

Fig. 15.7 (b) Metal: real vs. DK

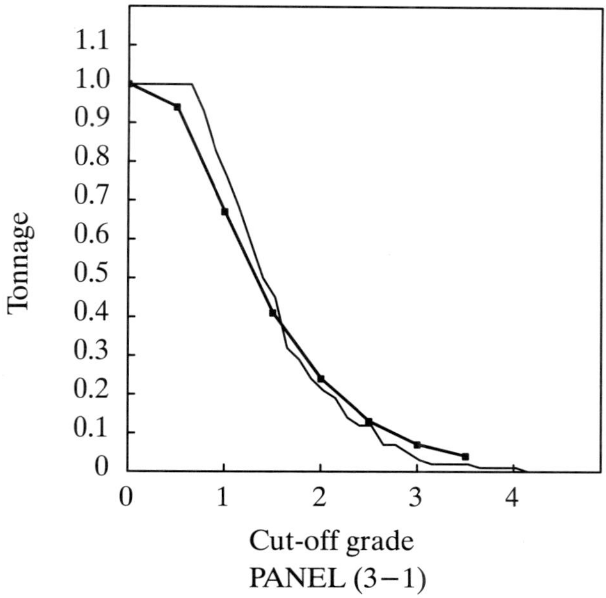

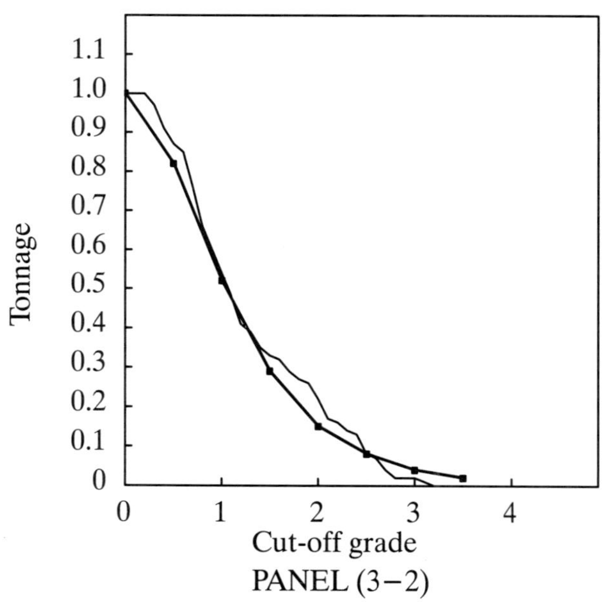

Fig. 15.8 (a, b) Tonnage above cut-off: comparison between real (thin continuous line) and estimate (thicker segments)

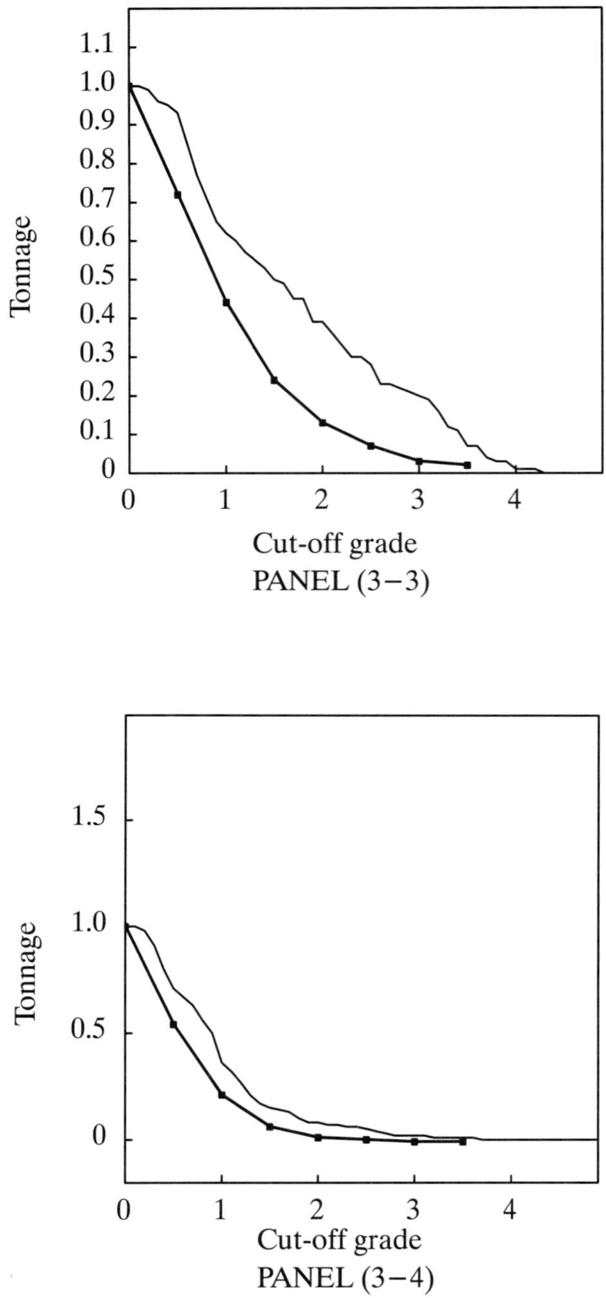

Fig. 15.8 (c, d) Tonnage above cut-off: comparison between real (thin continuous line) and estimate (thicker segments)

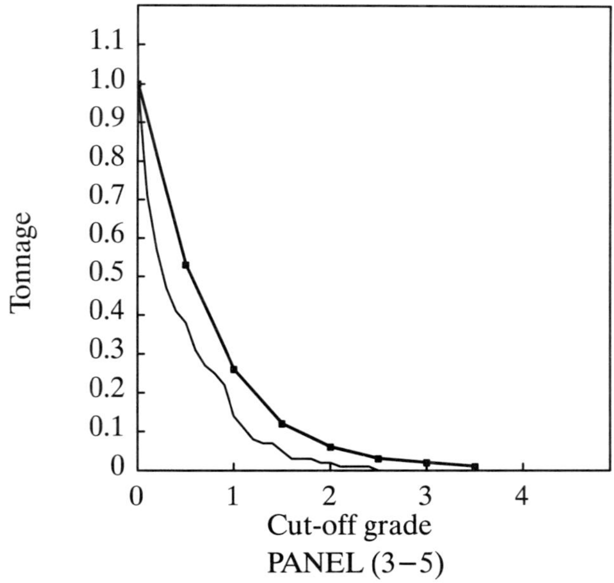

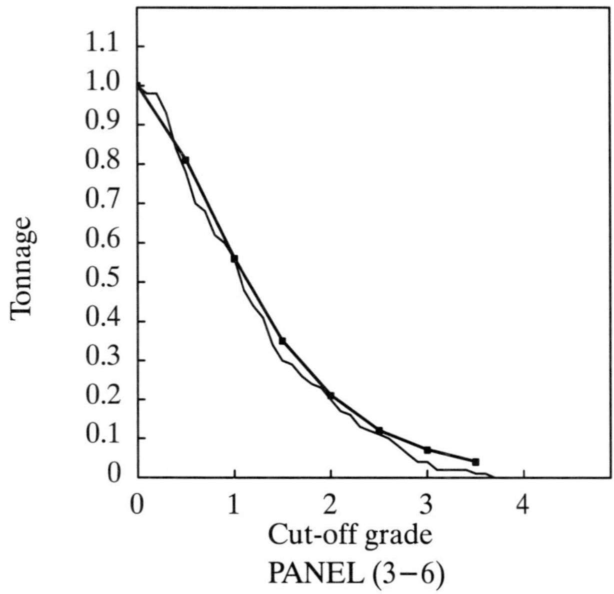

Fig. 15.8 (e, f) Tonnage above cut-off: comparison between real (thin continuous line) and estimate (thicker segments)

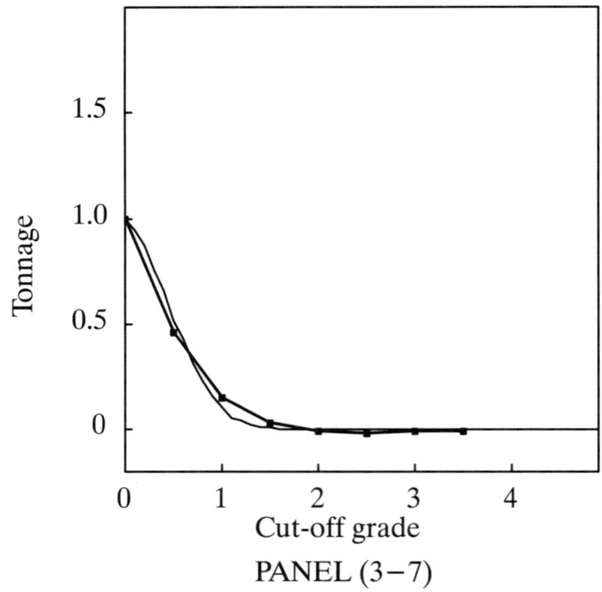

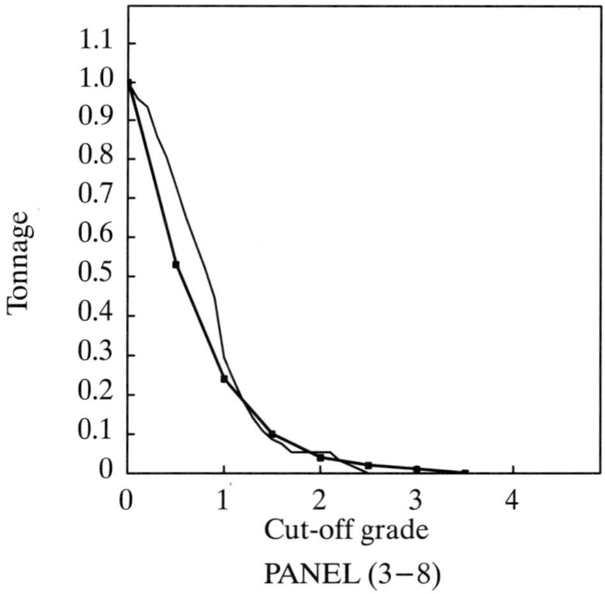

Fig. 15.8 (g, h) Tonnage above cut-off: comparison between real (thin continuous line) and estimate (thicker segments)

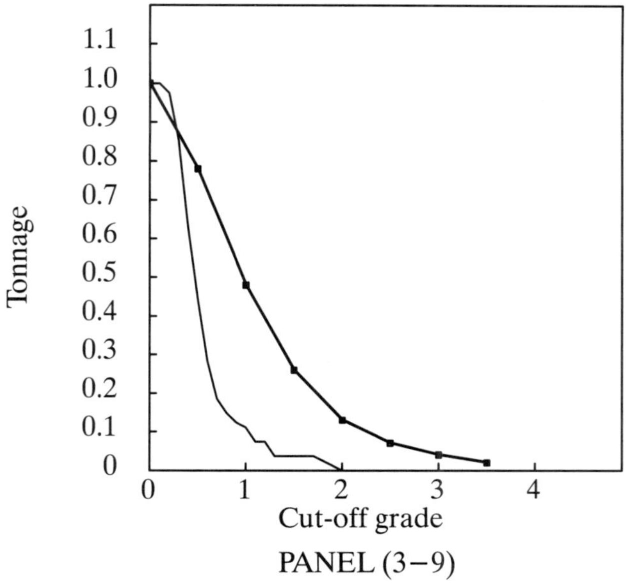

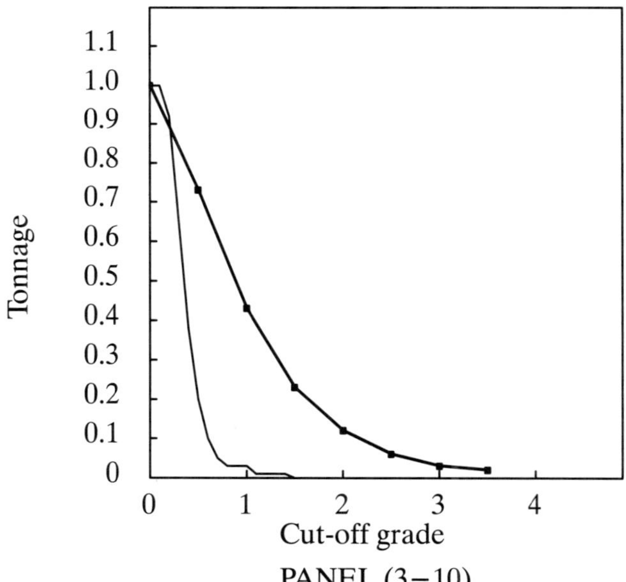

Fig. 15.8 (i, j) Tonnage above cut-off: comparison between real (thin continuous line) and estimate (thicker segments)

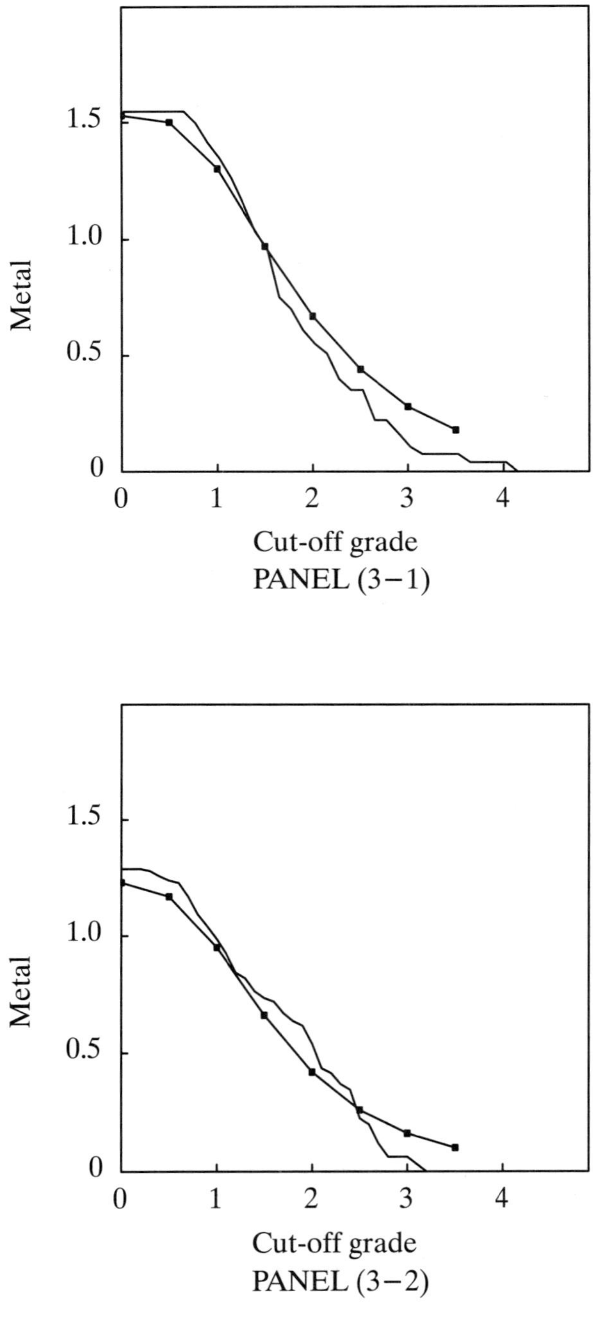

Fig. 15.9 (a, b) Metal above cut-off: comparison between real (thin continuous line) and estimate (thicker segments)

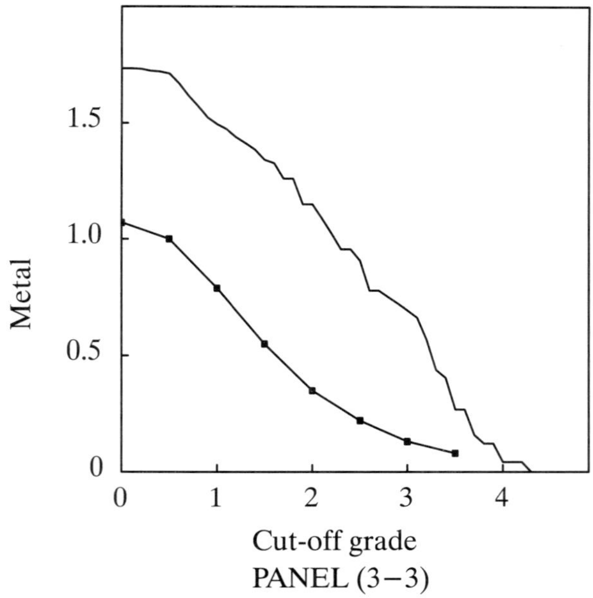

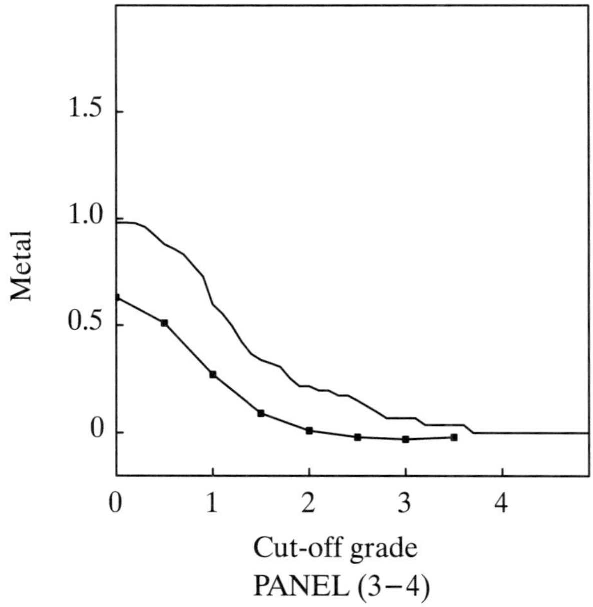

Fig. 15.9 (c, d) Metal above cut-off: comparison between real (thin continuous line) and estimate (thicker segments)

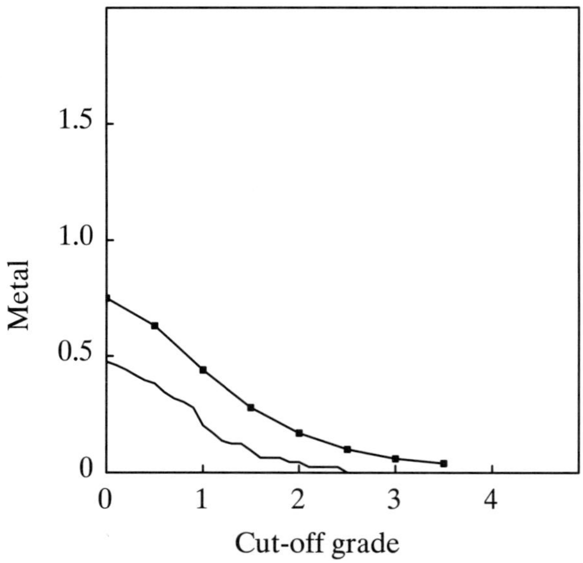

PANEL (3−5)

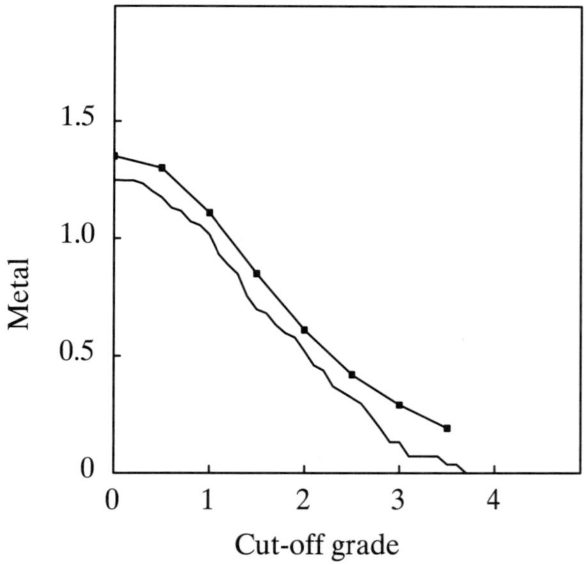

PANEL (3−6)

Fig. 15.9 (e, f) Metal above cut-off: comparison between real (thin continuous line) and estimate (thicker segments)

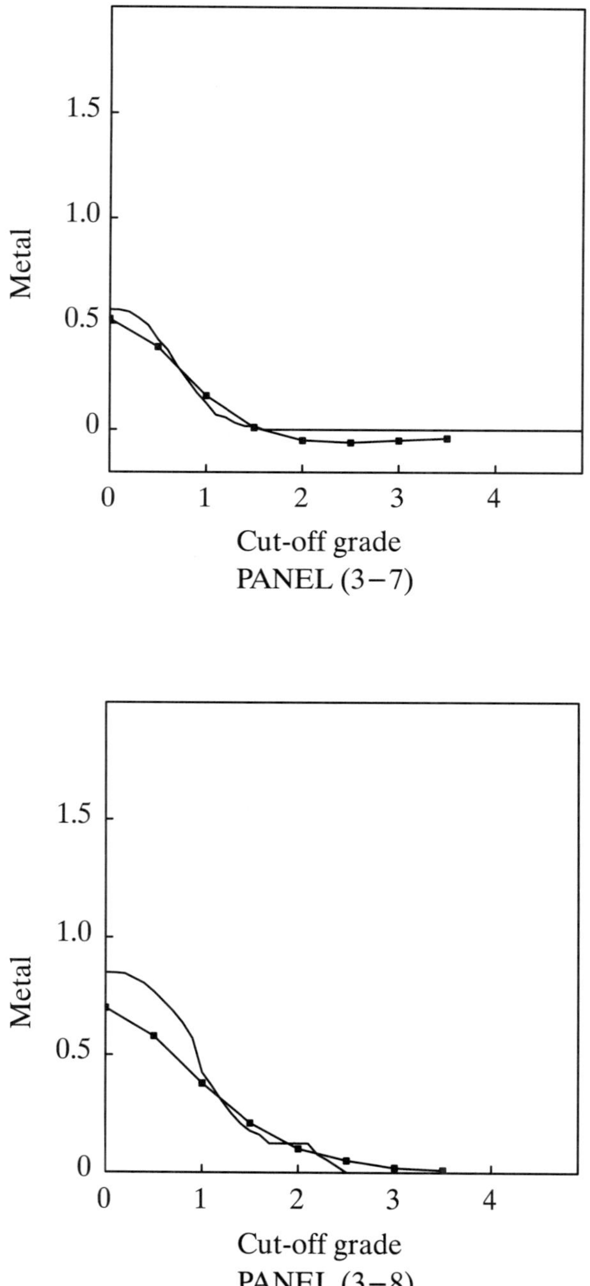

Fig. 15.9 (g, h) Metal above cut-off: comparison between real (thin continuous line) and estimate (thicker segments)

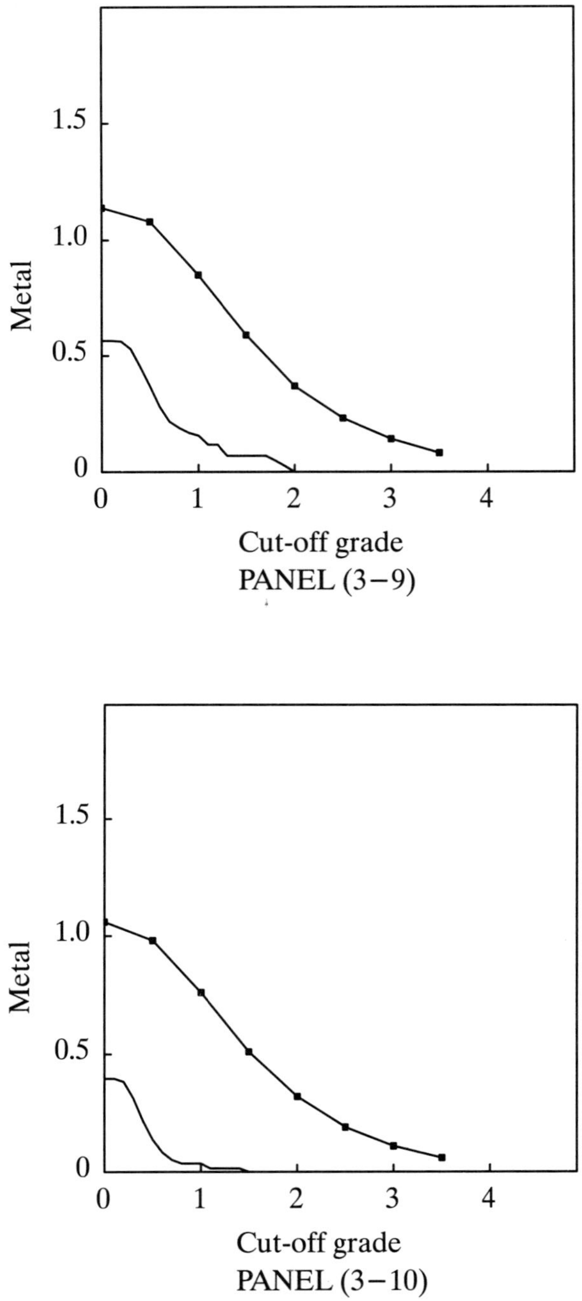

Fig. 15.9 (i, j) Metal above cut-off: comparison between real (thin continuous line) and estimate (thicker segments)

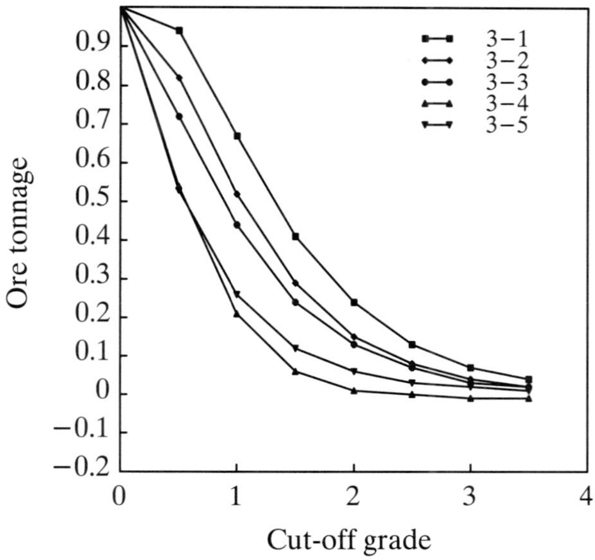

(a)

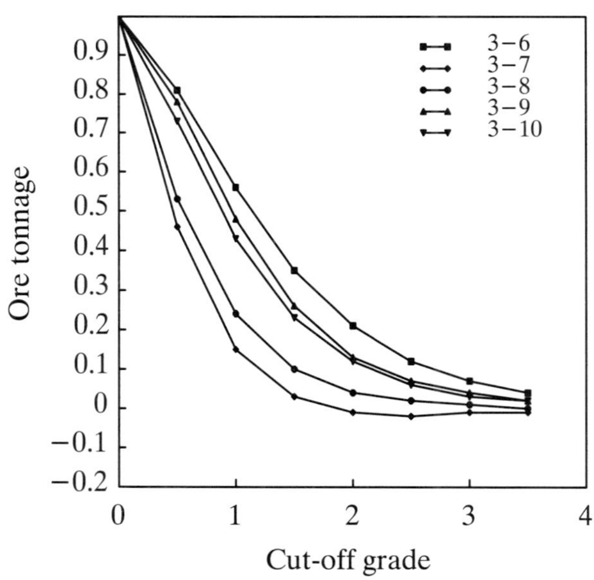

(b)

Fig. 15.10 Estimated ore tonnage as a function of cut-off grade

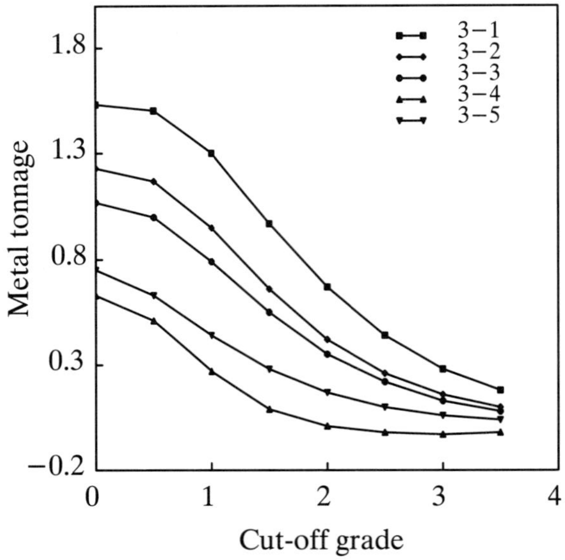

(a)

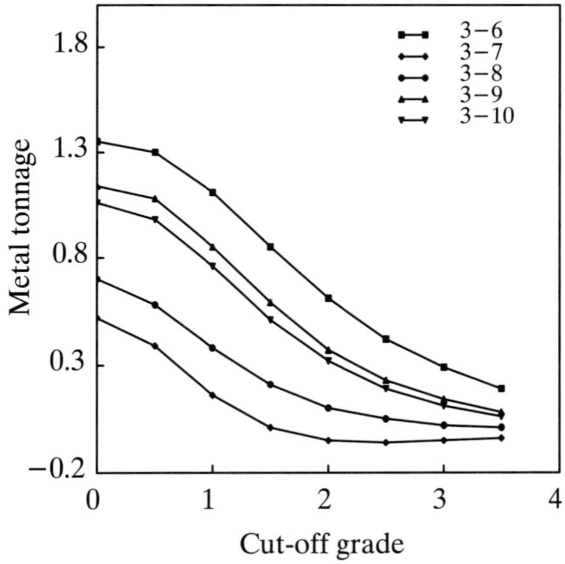

(b)

Fig. 15.11 Estimated metal as a function of cut-off grade

Index